Intelligent Industrial Internet Systems

Wanchun Dou • Xiaolong Xu • Shui Yu

Intelligent Industrial Internet Systems

Architecture, Deployment and Applications

Springer

Wanchun Dou
Department of Computer Science
and Technology
Nanjing University
Nanjing, Jiangsu, China

Xiaolong Xu
School of Software
Nanjing University of Information Science and
Technology
Nanjing, Jiangsu, China

Shui Yu
School of Computer Science
University of Technology Sydney
Ultimo, NSW, Australia

ISBN 978-981-99-5731-6   ISBN 978-981-99-5732-3 (eBook)
https://doi.org/10.1007/978-981-99-5732-3

This Springer imprint is published by the registered company Springer Nature Singapore Pte Ltd.
The registered company address is: 152 Beach Road, #21-01/04 Gateway East, Singapore 189721, Singapore

Paper in this product is recyclable.

Preface

In the process of transforming modern manufacturing into intelligent manufacturing, more and more manufacturing processes are being deployed digitally on the Internet. As a novel digital industrial infrastructure and new application ecology, Industrial Internet could integrate the novel information and communication technology into various industrial manufacturing processes, through the safe and reliable intelligent interconnection of factors concerning people, machines, and things. It aims at significantly improving the digitalization, networking, and intelligence of manufacturing processes through the comprehensive interconnection of all manufacturing factors. To deploy and run traditional manufacturing processes on Industrial Internet efficiently, corresponding digital technical means are called for. These technical means play an important role in many potential applications of intelligent manufacturing, for example, integrating product lifecycle data, strengthening industrial chain cooperation, providing decision-making service information oriented on the production organization process, and providing data support for product optimization and upgrading.

As an extension of cloud computing, mobile edge computing effectively promotes the popularity and application of IoT devices by providing additional computing resources with low latency. Thanks to the application of mobile edge computing and many other advanced technologies, the Industrial Internet has greatly improved the level of intelligence and constitutes an Intelligent Industrial Internet System together with these technologies. This book focuses on the theory, techniques, and means for integrating intelligent Industrial Internet applications deployed via mobile edge computing in this system. Concretely, it mainly studies the intelligent theory, techniques, and means from a problem-driven and demand-driven perspective, which supports collaborative design, agile manufacturing, remote operation, and maintenance on Industrial Internet. These technologies primarily include Industrial Internet-centric intelligent resource scheduling methods, various types of industrial data privacy protection, intelligent version management technologies supporting digital twin applications, various data mining algorithms for intelligent execution of business processes, etc.

Given its scope, the book offers a valuable theoretical and technical resource for project managers, technical developers, researchers, educators, and students at colleges and universities. The first chapter of this book provides a brief introduction to the background of Intelligent Industrial Internet Systems, allowing readers to gain a foundational understanding of the relevant knowledge covered in this book. The second chapter starts with the business processes management within Intelligent Industrial Internet Systems and describes its architecture along with several typical applications. The following chapters delve into detailed explanations of these applications, along with their corresponding algorithms and technologies. Readers can quickly grasp the content of each chapter through the abstracts, enabling them to find the specific technologies or applications they are interested in. It is our hope that readers will find this book insightful and rewarding as they engage with its contents.

Last but not least, we would like to extend our appreciation to the National Key Research and Development Program of China (Project No. 2020YFB1707601) for its support and funding. The main contents of this book reflect the works of our research and development initiated from the project.

Nanjing, Jiangsu, China Wanchun Dou
Nanjing, Jiangsu, China Xiaolong Xu
Ultimo, NSW, Australia Shui Yu

Acknowledgment

Writing a book is a labor of love that requires the collective effort and support of numerous individuals. As we bring this book to completion, we find ourselves overwhelmed with gratitude for the invaluable contributions and unwavering support we have received from so many exceptional individuals and organizations.

First and foremost, we extend our deepest appreciation to the National Key Research and Development Program of China (No. 2020YFB1707601) for their support and funding. Their belief in the importance of research and their commitment to advancing knowledge have provided the necessary resources and opportunities to undertake this ambitious project.

We are immensely grateful to doctoral students Xutong Jiang, Bowen Liu, Cheng Chen, and Yongzhe Jia for their crucial roles in this book. Their tireless efforts in material collection, meticulous writing, and diligent revision have been pivotal in shaping the content of each chapter. Their unwavering commitment to excellence and their intellectual prowess have elevated this work to new heights. Their insightful discussions, thought-provoking ideas, and thorough research have added depth and expertise, making this book a true testament to our collaborative spirit. We are deeply appreciative of their dedication and the countless hours they have devoted to this project. Their passion for knowledge and their unwavering support have been a source of inspiration throughout this entire journey.

We would like to express our sincere gratitude to Springer, the esteemed publisher of this book. Their commitment to scholarly excellence, rigorous editorial standards, and wide-reaching dissemination have provided me with an invaluable platform to share our research with the global academic community. We are truly honored to have partnered with Springer in bringing this work to fruition.

We would also like to extend our deepest appreciation to the editor of this book, Dr. Wei Zhu. Your meticulous attention to detail, astute editorial insights, and dedication to ensuring the highest standards of quality have significantly enhanced the overall coherence and readability of this book. Your guidance and expertise throughout the publication process have been invaluable, and we are immensely grateful for your unwavering support.

In conclusion, we would like to express our heartfelt appreciation to all those who have provided advice and support throughout the creation of this book. Your guidance, expertise, and encouragement have been instrumental in shaping and refining this work. We are deeply grateful for your contributions and the impact you have had on its development. Thank you for being an invaluable part of this journey.

Contents

About the Authors

Wanchun Dou is a full Professor of the Department of Computer Science and Technology, Nanjing University, P.R. China. His research interest includes IIOT, big data computing, Cloud-Edge Collaborative Computing. He has published two monographs and more than 200 technical papers at different venues, such as TPDS, TC, TIFS, TMC, TKDE, TETC, ToN, SIGCOMM, SIGMOD, and INFOCOM. Up to now, Professor Dou has chaired more than 10 national or state projects, respectively, supported by National Key Research and Development Program of China, National Natural Science Foundation of China, National Natural Science Foundation of Jiangsu Province, P.R. China, et al.

Shui Yu is a full Professor of the School of Computer Science, University of Technology Sydney, Australia. His research interest includes Cybersecurity, Network Science, Big Data, and Mathematical Modeling. He has published five monographs and edited two books, more than 500 technical papers at different venues, such as IEEE TDSC, TPDS, TC, TIFS, TMC, TKDE, TETC, ToN, and INFOCOM. His current h-index is 70. He is currently serving the editorial boards of IEEE Communications Surveys and Tutorials (Area Editor). He served as a Distinguished Lecturer of IEEE Communications Society (2018–2021). Now, he is a member of ACM and AAAS, and a Fellow of IEEE.

Xiaolong Xu (IEEE Senior Member) is currently a full professor at the School of Computer and Software, Nanjing University of Information Science and Technology, China. He worked as a researcher at the Department of Computer Science and Engineering, Michigan State University (2017–2018). He studied his PhD degree (2013–2016) in the State Key Laboratory for Novel Software Technology at Nanjing University. His research interest includes Cloud Computing, Big Data, Edge Computing, Deep Learning, and Federated Edge Learning. He has published over 100 peer-reviewed papers in international journals and conferences, such as TPDS, TKDE, TSC, ACM TOSN, and TITS. He was selected as the Highly Cited Researcher of Clarivate 2021 and 2022.

Chapter 1
Introduction

Abstract The Industrial Internet of Things (IIoT) is a subset of the broader IoT concept, which specifically emphasizes the integration of smart devices and automation technologies within industrial settings. IIoT represents a significant shift in how humans approach manufacturing, logistics, and supply chain management by leveraging real-time monitoring, automation, and data analytics.

One of the main advantages of IIoT is predictive maintenance, which involves the use of real-time data analytics and other machine learning algorithms to detect potential equipment failures before they occur. This approach can mitigate downtime, save costs, and increase efficiency. In addition, IIoT can facilitate remote monitoring, allowing for real-time tracking of production processes and equipment performance, enabling better decision-making, and reducing waste.

IIoT also has implications for supply chain management, by providing end-to-end visibility into supply chain processes, enabling greater tracking capabilities, and facilitating agile manufacturing approaches. This has the potential to significantly reduce inventory costs and improve customer satisfaction.

In summary, although IIoT plays a significant role in optimizing multiple aspects for organizations, the success of IIoT depends on thorough consideration of organizational needs, challenges, and goals. In the following sections, the book first introduces the resource scheduling problem in IIoT to achieve better resource utilization and reduce scheduling costs. Then the security enforcement and privacy-preserving in IIoT are discussed, and the final section introduces business process management.

The Industrial Internet is a more extensive concept than the Internet of Things (IoT) concept, which specifically emphasizes the integration of smart devices and automation technologies within industrial settings. The Industrial Internet represents a significant shift in how humans approach manufacturing, logistics, and supply chain management by leveraging real-time monitoring, automation, and data analytics.

The Industrial Internet can be viewed as a convergence of operational technology (OT) and information technology (IT). It combines the capabilities of sensors, actuators, programmable logic controllers (PLCs), and other automation technologies typically used in industrial environments, with cloud computing, big data

W. Dou et al., *Intelligent Industrial Internet Systems*,
https://doi.org/10.1007/978-981-99-5732-3_1

analytics, artificial intelligence, and machine learning. By integrating these technologies, the Industrial Internet enables companies to optimize their operations, costs, quality control, and safety.

With the integration of advanced technologies like edge computing, Artificial Intelligence and establishment of numerous infrastructures, the Industrial Internet has developed into a new form which we call Intelligent Industrial Internet Systems. They represent an advanced paradigm that builds on cloud computing and leverages mobile edge computing to enhance the integration and utilization of IoT devices with low-latency computing resources. By using mobile edge computing and other cutting-edge technologies, the Industrial Internet has witnessed significant advancements in intelligence, ultimately leading to the emergence of an Intelligent Industrial Internet System. This book focuses on exploring the theory, techniques, and methodologies involved in integrating intelligent Industrial Internet applications deployed through mobile edge computing within this system.

The primary objective of this research is to investigate intelligent theories, techniques, and means from a problem-driven and demand-driven perspective, enabling collaborative design, agile manufacturing, and remote operation and maintenance in the context of the Industrial Internet. Key areas of examination encompass Industrial Internet-centric intelligent resource scheduling methods, diverse industrial data privacy protection mechanisms, intelligent version management technologies supporting digital twin applications, and various data mining algorithms facilitating the intelligent execution of business processes.

In summary, although Intelligent Industrial Internet Systems play a significant role in optimizing multiple aspects for organizations, the success of Intelligent Industrial Internet System depends on thorough consideration of organizational needs, challenges, and goals. This chapter gives a general introduction to the background of the system. In the following chapters, the book first introduces business process management (BPM) within Intelligent Industrial Internet Systems and describes its architecture along with the introduction to several typical applications. Then the three chapters describe the resource scheduling problem in the system to achieve better resource utilization and reduce scheduling costs. The sixth to eighth chapters characterize security enforcement and privacy-preserving in the system, and the final three chapters introduce several applications and algorithms in BPM over Intelligent Industrial Internet Systems.

1.1 Industrial Internet-Centric Intelligent Resources Scheduling

This section delves into the challenges [1] associated with utilizing cloud computing for IoT, particularly the latency issues that arise, and introduces edge computing as a viable alternative. An edge-enabled IoT service orchestration (EISO) framework is introduced to optimize resource allocation based on service delivery rate, delay, and latency constraints. The framework consists of five layers and is scalable and energy-efficient in real-world scenarios.

Aiming at the challenge of IoT service orchestration caused by different service latency requirements, variable resource demands, and other diversified requirements, this chapter presents a new EISO framework that enables edge computing for service provisioning, ensuring optimal delivery rate and latency. The proposed framework replaces central cloud computing with computation resources near the IoT's edge to reduce latency during remote process execution. The optimized resource allocation based on service delivery rate and latency guarantees efficient service provisioning by performing analytics, stream mining, and data filtering. An analytical framework is formulated to solve the optimization problem, followed by the discussion of design parameters in five layers of' service provisioning framework, including perception, mist, edge, cloud, and application. The performance evaluation shows that the proposed framework efficiently allocates resources, offering a feasible solution for IoT service provisioning under diverse constraints.

1.1.1 Related Work

The section discusses various techniques for QoS-aware service provisioning in IoT using edge computing, with a focus on resource provisioning efficiency, resource provisioning at the edge, heterogeneous latency services, and compute-intensive tasks on the edge-cloud paradigm. The authors organize literature into multiple categories and provide an overview of relevant studies, while also highlighting the need for optimal resource provisioning strategies that strictly follow latency and delivery-rate requirements. Furthermore, they propose a framework for efficient resource allocation in the hybrid edge-cloud paradigm based on delay and service delivery rate for different latency services. IoT is a dynamic framework that emphasizes ubiquitous real-time service provisioning, management, and storage. In the literature, various QoS-aware service provisioning techniques have been introduced, but this section will focus on relevant studies and demonstrate the efficacy of the proposed framework. The state-of-the-art literature on service provisioning supported by edge computing has been organized into multiple categories, which are discussed in the following sections.

Enhancing Resource Provisioning Efficiency: Resource provisioning efficiency in IoT can be enhanced by relying on edge computing for resource allocation to compute-intensive tasks. Various publications [2–6] have proposed resource provisioning in IoT using edge computing, such as tensor representation, incentivized strategy, mist computing, fog computing based on container technology, and fog computing for latency-sensitive IoT applications. However, optimal resource provisioning strategies are still needed while strictly following latency and delivery-rate requirements.

Resource Provisioning at the Edge: Extending cloud resources at the edge in IoT also improves service provisioning efficiency and performance [7–

11]. Examples include fog computing frameworks, rate-adaptive fog resource platforms, and QoS-aware frameworks for smart healthcare or heterogeneous IoT. These approaches lack mist computing, which can reduce extra resource consumption and network overhead.

Heterogeneous Latency Services: Edge analytics can reduce the computational overhead on the cloud and save extra network bandwidth consumption [12–16]. Edge analytics approaches for IoT include distributed edge nodes, deep learning-based frameworks, end-edge-cloud frameworks, predictive modeling frameworks, and vehicular communication techniques. However, complex computations may cause delays in data transfer, inducing additional overhead in resource provisioning.

Compute-Intensive Tasks on Edge-Cloud Paradigm: Compute-intensive tasks are crucial for deriving knowledge and delivering vital information for controlling real-world infrastructures in IoT services [17–20]. Analytics frameworks for data-driven services, layered frameworks for processing compute-intensive tasks in mobility-constrained edge nodes, and big-data sharing frameworks have been proposed. However, these approaches may consume more computational resources, require large historical datasets, or encounter extra delays in the chaining process of blockchain.

Although various techniques for resource provisioning have been proposed in edge computing, there is still a lack of optimal resource allocation research that ensures higher performance by optimally allocating resources to requesting nodes based on delay and service delivery rate for different latency services in the hybrid edge-cloud paradigm. The following section will introduce a framework for efficient resource allocation.

1.1.2 Edge-Enabled IoT Service Orchestration Framework

Large-scale IoT deployment generates huge amounts of data with limited resources, which requires efficient and QoS-aware processing. Cloud computing alone is inefficient and causes increased network resource consumption and poor performance. Processing at the "edge" of IoT enhances data processing, reduces response time, and improves network resource utilization. Mist computing integrates virtual computing resources with IoT devices, increasing performance and reducing response time. Data can be filtered at the edge of IoT based on specific conditions while multi-layer computation reduces network overhead. The EISO framework encompasses multiple layers, including perception, mist, edge computing, central cloud computing, and application layers, all of which work together to provide improved resource utilization and enhanced quality of service (QoS) efficiency.

The layered framework has been shown in Fig. 1.1 representing five layers. Seamless coordination among all layers is necessary to provide QoS-aware services from the source to the destination. IoT allows for the analysis of multiple activities in the IoT deployment paradigm such as smart cities. Edge computing-based solutions

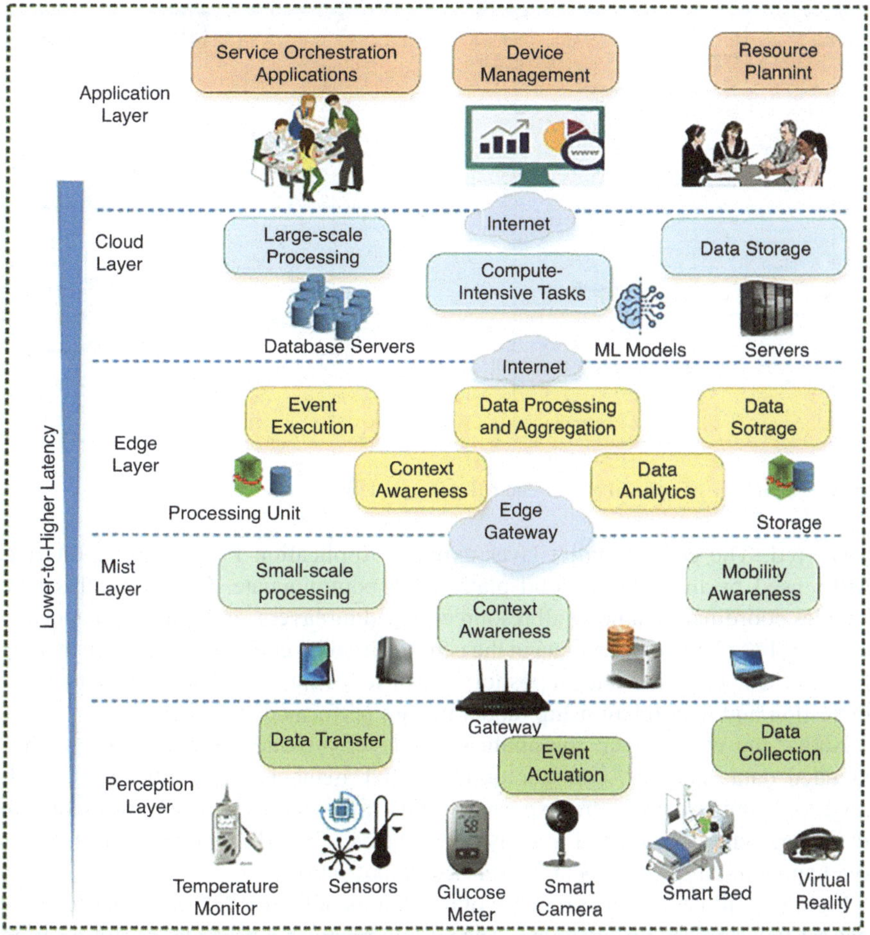

Fig. 1.1 Architecture of the EISO framework

offer effective latency-sensitive job scheduling for traffic monitoring using sensory nodes deployed in smart cities.

The perception layer captures physical and digital data using low-cost sensory nodes placed in critical areas to monitor and control infrastructures.

The mist layer enables computational deployment, management, and monitoring of sensors, taking care of situational data and mobility management.

The edge computing layer brings computing resources closer to users, comprising low and high computational power nodes distributed around resource-constrained IoT devices. It integrates latency-critical computations and forwards latency-tolerant computations to the central cloud for further processing.

The cloud layer, on the other hand, offers localized computation for low-latency services. It acts as a centralized data center where compute-intensive tasks can be

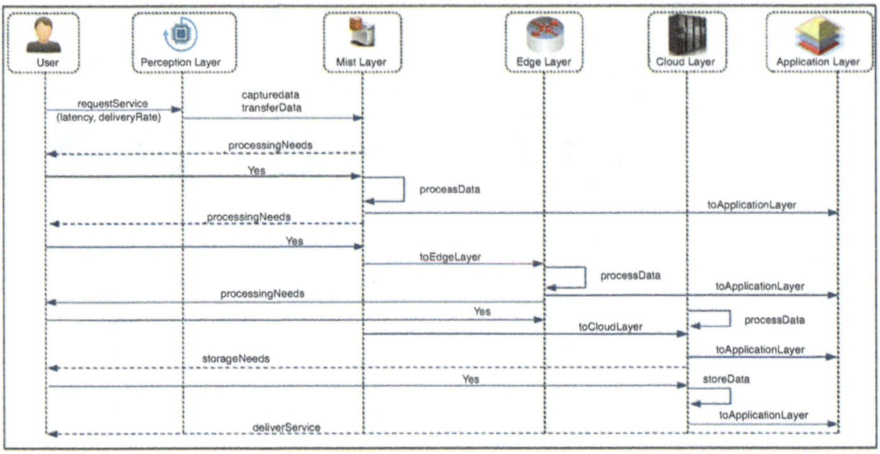

Fig. 1.2 Sequence of optimal end-to-end IoT services delivery

transferred. The edge cloudlet layer exposes Application Programming Interface (API)s used by the central cloud platform to communicate. Moreover, this layer provides coordination among all resource-providing layers, and long-term decision-making is based on archived data in the central cloud layer. Complex pattern analysis based on a centralized data repository, such as event detection related to traffic congestion and data-relationship modeling, are performed at this level.

Above the cloud layer, the applications layer provides core services, such as extended data analytics, configuration, and IoT management applications, to the service consumers. APIs are exposed by this layer, along with an abstraction for developers who do not need to consider complex network parameters, to facilitate novel application development for service orchestration.

Figure 1.2 illustrates the sequence of operations of the EISO framework for edge-enabled IoT services provisioning. The system receives a user request for a service based on a specific delivery rate and latency. Real-world data is captured by the perception layer, processed by the mist layer at the extreme edge to ascertain further processing needs, and transferred to the edge layer for the required processing. Results requiring archival are stored in the cloud layer, while others are transferred directly to the applications layer. For higher latency services, computation is performed by the central cloud, which archives the data for future use. This seamless collaboration among all layers efficiently offers end-to-end services from service providers to consumers. The next section will cover the analytical framework for service orchestration based on service latency and delivery rate parameters.

1.1.3 Problem Formulation for Optimal Resource Allocation

This section discusses service provisioning in the hybrid cloud paradigm for supporting elastic and real-time latency-sensitive services. To allocate efficient resources to consumers based on service delivery rate, time delay, and packet drop rate, this section devises an analytical framework for optimal service delivery rate and formulates latency-aware resource allocation using a hybrid cloud paradigm. This involves first identifying the optimal service delivery rate for a service and then targeting service provisioning resources based on latency requirements.

The network consists of end IoT devices and edge computing elements connected through links represented by the set. Services are defined as a set of predefined tasks. The service consumers/applications consume these services, and a source-to-destination service delivery tree is established between them. Resource providers are located at the root node, while the service receivers are at the leaf nodes connected via the perception layer.

For instance, the QoS-aware service delivery rate for IoT devices/applications from the edge computing service provider group is denoted by a graph, which is the set of paths from service provider to consumer. The per-service resource allocation efficiency can be evaluated by a positive, increasing, and continuous function within a certain range. This utility function must be strictly positive and concave for both latency-sensitive and latency-tolerant services. The goal is to maximize resource allocation efficiency while considering delay and packet drop rate requirements. This can be formulated as a convex optimization problem to maximize overall system utility. The evaluation ensures that all user data processing occurs at the mist/edge/cloud layer and that the total amount of data received from consumers equals the final processed data at these nodes.

To maximize service delivery based on the constraints, the framework derives a Lagrangian-constrained optimization problem and has a multivariable function that needs to be maximized. Each service provider incurs a cost in terms of resource consumption, energy, and price associated with the link they use. The Lagrangian multiplier satisfies to maintain the value. Faster receivers tend to have increasing values, whereas slower receivers have decreasing values.

The framework categorizes the problem into three classes: latency-sensitive (LS), latency-tolerant (LT), and mixed (M) services based on the data transmission rate and queuing delay. The goal is to reduce overall delay and packet drop rate during service orchestration while considering user requirements for time delay, packet drop rate, bandwidth demand, and buffer length demand of IoT devices. Then calculate resource allocation based on IoT device requirements. To minimize overall delay, the framework uses the M/D/I queue model. Here, a M/D/I queue represents the queue length in a systems has I servers, where arrivals are determined by a Poisson process and job services times are fixed. Packet drop occurs when the required average queue length exceeds the demand buffer length size. Link distribution and aggregation techniques are used to reduce end-to-end delay, where multiple links are selected and the traffic is optimally distributed among the links.

The framework focuses on flow control and advocates for a relationship among QoS performance of services based on delay, service delivery rate, and packet drop rate requirements. Service delivery is optimal when service providers have sufficient resources. Admission control is necessary when resources are not available and resource providers are busy. However, real-time services may be exempted from this condition for efficient service delivery. The IoT system controller coordinates with links to distribute the load based on delay-sensitive services and fixed services demand requirements. Resource requirements should not exceed the available resources in terms of space and computation constraints.

1.2 Industrial Internet-Centric Privacy and Security

1.2.1 Security-Enforcement Technique

The rapid deployment of edge-enabled IoT devices for service orchestration has increased their vulnerability to DDoS (Distributed Denial-of-Service) and flooding attacks. Due to the limited memory of IoT devices, implementing device-specific security solutions is challenging, leaving them vulnerable to various types of attacks. As a result, the network connectivity of these vulnerable devices exposes the global network to attacks, ultimately resulting in the saturation of limited resources. In addition, users often rely on default credentials, making it easier for attackers to gain access to IoT devices and launch large-scale attacks, such as the Mirai botnet attack.

To address these vulnerabilities, this section introduces a network-level security solution using SDN characteristics and extending the EISO framework to provide defense against attacks on resource-constrained IoT devices. The presented solution leverages network traffic metrics to identify attacks and employ traffic engineering techniques to secure the network and allow only legitimate traffic. By securing the data plane of the extended EISO framework, seamless end-to-end IoT service orchestration can be achieved. Figure 1.3 depicts the architecture of the extended EISO framework, connecting IoT devices with SDN switches at the data plane.

1.2.1.1 Securing Edge-Enabled from Attacks

This section addresses vulnerabilities in the data plane of the extended EISO framework for edge-enabled IoT service orchestration as IoT devices are resource-constrained and can be easily attacked by DDoS. This framework focuses on addressing the integral component of IoT security to ensure efficient end-to-end service delivery. The resource limitations of IoT devices make them vulnerable to various attacks, including DDoS attacks, which are among the most devastating against current data center networks. The adversary can access IoT devices and program them to send attack packets, probe the victim's network, and attack data plane switches with flood packets to cause packet misses and trigger new flow rule installation. These vulnerabilities present in the data plane of the EISO framework could be exploited by adversaries to attack the network infrastructure.

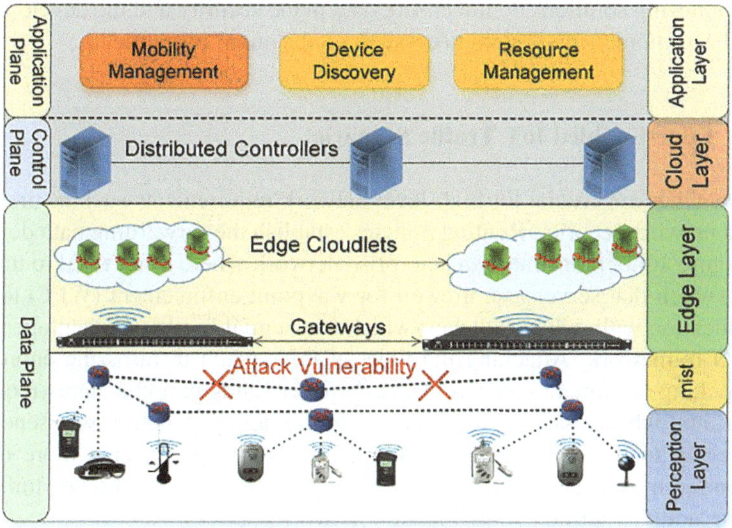

Fig. 1.3 The extended EISO framework

1.2.1.2 Vulnerabilities of Edge-Enabled IoT

The performance of the entire network can be affected by data plane attacks that interrupt the packet flow among connected IoT devices and may result in IoT service outages, communication disruption, information loss, and increased operational costs. In SDN update processes, per packet consistency (PPC) is the enforced policy to maintain high-level network requirements. Waypoint enforcement (WPE) diverts network traffic to checkpoints, such as "policy switches," which are essential for IoT traffic to traverse and ensure a high level of control and security. The data plane implements packet rule implementation, including traffic forwarding, policy enforcement, and security, making it susceptible to different types of attacks, including DDoS and crossfire attacks (CFAs), launched by smart adversaries on critical switches to disrupt communication.

To enable seamless service orchestration, securing the data plane is necessary due to the widespread security challenges faced by the EISO framework. Crossfire attacks (CFAs) present a significant threat as they inundate crucial links, leading to the disruption of communication with critical servers. These attacks employ rate-limiting techniques and generate traffic that is indistinguishable from legitimate flows. Consequently, traditional flow filtering and packet inspection techniques become ineffective since attackers do not inject malicious information into the packets but instead employ low-rate flows that appear legitimate. Resource-constrained IoT devices make it infeasible to deploy security solutions, rendering them more susceptible to CFAs. In addition, these devices are vulnerable to being manipulated to attack vital communication infrastructure, making defending CFA on the data plane infrastructure in edge-enabled IoT challenging. Currently, control plane monitoring solutions do not ensure data plane security and the development of a specialized monitoring system necessitates substantial investments.

1.2.1.3 Edge-Enabled IoT Traffic Scenario

Policy switches are crucial for high-level network requirements such as firewalls to maintain network security. Routing policies establish the forwarding path demand in SDN. Figure 1.4 demonstrates an enterprise network where flows need to traverse a specific switch that serves as a firewall for waypoint enforcement (WPE) to ensure communication with policy switches, which is essential for the efficient operation of the EISO framework. Attacking the links of this switch disturbs the entire traffic path, blocking legitimate services. The controller sends periodic status requests to switches, and when a status request is sent to the policy switch, it will send back a normal reply to the controller, allowing the adversary to execute more damage. Links around the policy switch continue flooding, preventing legitimate traffic from traversing and disrupting communication network services.

Assuming that only connecting the access/policy switch to the controller designates it as the access provider, which enhances the efficiency of service delivery. Coordinated bot attacks with low-rate traffic exhaust target link bandwidth without exceeding detection limits. Attackers compromise IoT devices using firmware updates and client certificates installed on edge cloudlets to access network resources and launch low-rate CFA toward policy switches. This framework focuses on degrading performance in an in-band controller paradigm that isolates the link to the policy switch from the rest of the network to eventually bring down the entire network. Here we conduct an attack scenario where 5 IoT bots and 3 decoys flooded the policy switch links in the data plane. With 8 Mbps traffic from each IoT bot, the attack ensures the traversal of the policy switch. The remaining bandwidth decreases continuously as the attack rate rises until it reaches 2 Mbps at an attack rate of 25 Mbps.

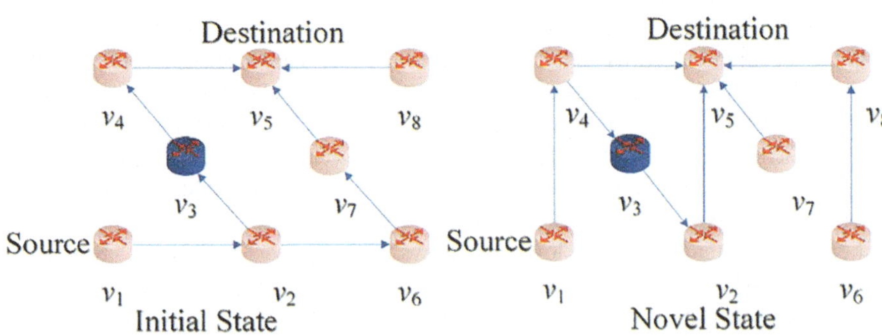

Fig. 1.4 A WPE scenario

Round trip time (RTT) will increase as the attack rate rises, with RTT reaching 4 ms at a 20 Mbps attack rate and rising sharply at 25 Mbps, indicating degradation in link status. At 25 Mbps, available bandwidth decreased significantly while RTT increased, causing the link status to change from normal to flooding. Low-rate traffic was used to flood policy switch links, leading to these effects. When traffic reaches 25 Mbps, expected packets are not received at the destination due to flooded links. Deploying low-rate traffic could obstruct legitimate flows by disrupting services to the policy switch. This reinforces the lethality of CFAs on policy switches as they can degrade network performance and disrupt service provisioning in edge-enabled IoT.

1.2.1.4 System Modeling and Analysis

Attacks on important switches can disrupt overall security, damage the PPC, and cause source-to-destination delivery anomalies and other inconsistencies. Existing countermeasures against crossfire attacks (CFAs) primarily focus on securing the controller, switch memory, or OpenFlow channel, but they often prove ineffective when it comes to protecting policy switches within software-defined networking (SDN) environments. Most techniques discussed to counter CFA are either developed for traditional networks or focus on detecting or increasing network connectivity in SDN.

Low-rate attack traffic floods the policy switch links to show the effect of such attacks on the EISO framework by inducing high delays and severe congestion. This section introduces a model, as shown in Fig. 1.5, to deal with this problem; This model serves as a lightweight, application-level countermeasure for the EISO framework and comprises three modules: link selection, attack detection, and malicious flow interception modules.

The model is a technique developed to detect and mitigate crossfire attacks (CFAs) on the data plane of the EISO framework. Traditional authentication mechanisms fail due to the use of low-rate legitimate attack traffic, and defense mechanisms like path and IP inspection fall short of the distributed nature of IoT botnets. To defend against attacks, the model utilizes centralized control characteristics of the EISO framework. It consists of three modules: link selection, attack detection, and a malicious flow interception module, each having subcomponents for seamless operation. The link selection module identifies target switch links before the attack by analyzing data plane devices. The attack detection module uses traffic engineering techniques to identify CFA. Finally, the flow interception module mitigates the attack traffic by providing bypass links, traffic rerouting, and blacklist elimination. The controller analyzes the flow status on all other links during the attack detection and inspects link congestion, changing the policy switch status to another switch. Target links are selected based on the highest flow weight, and attack detection and malicious flow interception modules alleviate network congestion and identify the attack.

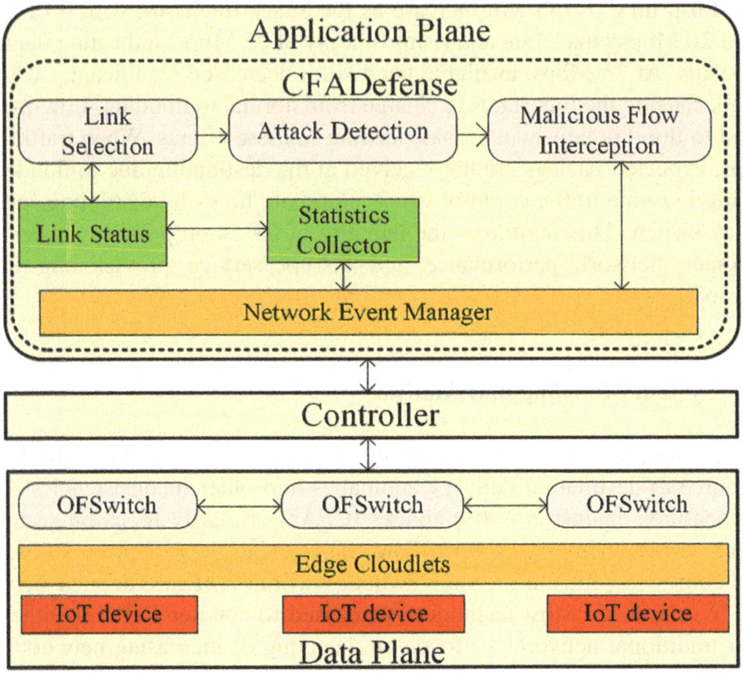

Fig. 1.5 CFADefense components

The model uses the reactive applications approach of the floodlight controller to receive notifications, with three listeners: switch, device, and message listener. The malicious flow interception module alleviates congestion by allowing traffic to different paths and installing drop rules. It selects links with higher flow density during traffic analysis and detects congestion with the REST API of the floodlight controller, blocking malicious flows through the interception module.

Link Selection Module

Attackers often target links with high flow density to cause maximum damage to the network. In CFAs, attackers create a map of precise links to identify potential links for attacks and understand the network topology. The model uses the switch listener of the floodlight controller to extract switch status information. The flow path from source to destination is computed by analyzing the next-hop information of the ingress switch. The switch-link tuple is initialized as NULL before examining the links and calculating the number of flows passing through them between all pairs of switches in the network. The target link selection module then iteratively traverses the next switch, analyzes its flow table, and computes the flow weight of the link again. Finally, it increments the weight of each link, selects the link with the highest weight after repeating the process on all switches, and computes the number of flows on all the links.

Attack Detection Module

SDN-enabled IoT service orchestration networks currently lack advanced congestion detection capability. Therefore, the model incorporates an attack detection module into the extended EISO framework's centralized network management characteristics to address this issue. The attack detection module has two sub-components: link inspection and attack detection modules. The link inspection module analyzes flow statistics to calculate packet loss rate, jitter, RTT, and throughput. However, the attack detection component identifies attacks by monitoring changes in these statistics. When a higher packet drop rate is observed, the model keeps track of network traffic and employs network traffic capturing strategies with the message and device listener modules of the floodlight controller. This helps to detect attacks and notify the malicious flow interception module for action.

The model employs a congestion detection module that measures traffic flows and identifies congestion with a threshold value. While querying switches for flow rate may not always yield accurate values, a significant decrease in flow rate is indicative of congestion. When congestion is detected, an alarm is triggered, and the switch sends information such as the data path ID (DPID), flow stats, and port ID to the controller. With knowledge of the forwarding path of all flows, possible attack nodes can be identified. A hop-by-hop measurement is also implemented to infer flooded links. If congestion occurs between two switches, the detection module queries the controller to install two flow entries with match fields, counter, priority, instructions, and timeout. The match fields specify Internet Protocol (IP) and Internet Control Message Protocol (ICMP) values and the destination of the first flow rule is installed on a specific switch, with forwarding assigned to the appropriate port. The controller subsequently sends an ICMP packet with a time-to-live (TTL) value of 1 toward the next switch. If an ICMP-unreachable message is received, the precise location of congestion is determined. If instead, the switch receives an echo message, it analyzes the second entry in the flow and transfers the packet toward the controller, which further analyzes it. If necessary, the controller sends another ICMP packet with a TTL value of 1 toward the destination switch.

Malicious Flow Interception Module

This module employs three strategies to eliminate flooding flows: traffic rerouting, blacklist elimination, and switch role change. Traffic rerouting analyzes network congestion and ascertains the shortest optimized route toward the destination. Malicious flows are identified during the rerouting process. Constant flooding flows by adversaries owning limited bots are detected when they send traffic toward a high-density flow area. The model then blocks these malicious flows using a Flow

Set. Blacklisting helps alleviate severe link congestion by dropping traffic from malicious hosts. In the process, the hostname of the received packets is first extracted. If the extracted hostname matches a host listed in the malicious database, the packet is subsequently dropped. This mechanism allows for the identification and prevention of potentially harmful network traffic from reaching its intended destination. During network congestion, top-N malicious flows in the database are also dropped to normalize congestion. Bypass links are provided to alleviate flooding and reroute traffic toward another switch. Switch role change forces adversaries to change their target while traffic rerouting and malicious flow interception run simultaneously, effectively eliminating the attack.

1.2.2 Privacy-Preserving Technique

IoT is a promising technology with applications in smart cities, homes, transportation, and industrial control systems. However, resource-constrained IoT devices lack the computation resources to process generated data, resulting in reliance on cloud infrastructure for decision-making and transfer, which creates privacy vulnerabilities. This section introduces how to achieve privacy-preserving by utilizing centralized network management through an SDN-managed EISO framework and distributed blockchain technology to address the privacy challenges.

1.2.2.1 Related Work and Motivation Analysis

IoT data is often used for service provisioning using physical systems, creating ubiquitous and autonomous systems. Edge computing extends cloud resources to the edges of IoT, but requires continuous interaction between IoT and edge, highlighting security and privacy issues due to insufficient authentication. Traditional techniques rely on centrally managed policies, resulting in single-point-of-failures and trust deficits which cause privacy issues such as information leakage and identity theft. Thus, privacy preservation in edge-enabled IoT is crucial for effective service provisioning. IoT users are frequently unaware of secure smart object usage, leading to privacy leaks and potential attacks. Such leaks can cause revenue loss, operational disruption, and legal ramifications. Privacy threats arise from distributed, resource-limited, and large-scale IoT infrastructure that traditional centralized approaches overlook, making IoT vulnerable to privacy attacks that compromise personal data. Protecting privacy is critical given the potentially lethal consequences for organizations. In summary, this passage discusses the importance of privacy preservation in IoT and related work motivations.

Verifiable computation [21] is a technique that leverages compute nodes to offload a task. In this process, an IoT device generates a public encryption key, and both the key value and computation can be authenticated using integrity evidence obtained from edge devices. This approach ensures the validity and integrity of the computation performed, enhancing the security and trustworthiness of the overall system.

Currently, the following techniques can be utilized to enhance privacy-preserving. EdgeChain prototype [22] uses a secure blockchain currency/coin system to link edge cloudlets with IoT devices and create a secure ecosystem for service provisioning and orchestration. A secure cloud data storage approach [23] where only authenticated devices store data using distributed mechanisms on the blockchain. An AI-based hash is utilized by the secure smart city management framework for authenticated transaction storage on the edge. However, incorporating artificial intelligence services for data processing may hamper the overall system performance, and differentiating compute-intensive tasks from non-compute-intensive tasks can add more overhead to the system.

The transactions in the edge-enabled IoT service provisioning paradigm require authentication between entities. Agreement-based decision-making using session keys is a widely used authentication technique [24, 25], but it increases communication overhead and complexity for resource-constrained IoT devices. An authenticated key agreement (AKA) protocol generates common session keys and creates a consensus between entities. However, AKA suffers from deployment challenges over multiple edge cloudlets. Blockchain eliminates the need for centralized security mechanisms and provides a distributed consensus protocol to support trustworthy interaction among unknown devices with its immutable public ledger. Blockchain ensures the authenticity, reliability, and integrity of distributed IoT devices by eliminating single points of failure. The blockchain records are safe from unauthorized access with only authenticated nodes having access to resources. A new entry in the blockchain is stored on top of the previous blocks, and any malicious activity can be easily detected since all nodes store the same set of data. To address privacy challenges and attack vulnerabilities in IoT, this study proposes a trust-aware system among edge and IoT nodes using blockchain, offering edge cloudlets a secure environment for accessing network resources and ensuring overall network privacy. An algorithm using a public/private key consensus to authenticate IoT and edge devices, which are deployed on the edge cloudlets for IoT device authentication.

1.2.2.2 System Modeling and Analysis

The existing research on privacy protection in IoT using blockchain has limitations and tends to focus on specific aspects of communication, rather than considering the whole IoT ecosystem. For example, in the smart vehicles paradigm, interactions with other smart capabilities such as hospitals, weather forecasting systems, road signals, and refueling stations increase the risk of privacy leakage, which has not been fully explored in previous studies. Therefore, a decentralized authentication mechanism based on blockchain is introduced as an efficient solution for addressing privacy concerns in edge-enabled IoT. The mechanism is lightweight, secure, and scalable for large-scale network requirements.

The mechanism utilizes blockchain technology, which can be divided into public and private types. A public blockchain is a peer-to-peer (P2P) decentralized system with no private ownership where participants work toward a common goal through consensus. Private blockchain, on the other hand, is more centralized and has agreed-upon regulations to operate, offering users extended privacy. In blockchain architecture, each participant approves, maintains, and updates entries, enhancing network security. Blockchain nodes are endpoint workstations with a complete copy of the database updated independently, while transactions are records in the blockchain stored in blocks verified by miners before being added. Every block contains a block-hash, generated using cryptographic algorithms like SHA-256, which is automatically added when a block is created to maintain data integrity. A malicious attempt to alter a block leads to incorrect information in subsequent blocks, rendering the entire blockchain invalid. Every joining node receives a copy of the blockchain system to synchronize with the new blockchain status. The Elliptic Curve Digital Signature Algorithm (ECDSA) uses the elliptic curve and a finite field to create and verify a signature between parties.

This section highlights the potential privacy-spoofing attack scenario in the presented IoT solution due to the lack of computational resources. The attackers can exploit various vulnerabilities, including man-in-the-middle (MiTM), Link Fabrication Attack(LFA), DDoS, and privacy attacks, by taking advantage of the resource-constrained nature of IoT devices. Autonomous transactions in IoT make it easier for adversaries to infiltrate transaction systems and cause data and revenue losses through fraudulent financial transactions without revealing their identity. Moreover, attacks on autonomous systems can lead to severe consequences such as human life threats, identity theft, revenue, and information loss. This section discusses the blockchain's working principle, adversary model, and attack vulnerabilities, leading up to the introduction of the solution's components aimed at preserving IoT privacy.

Components

The algorithm utilizes blocks in the blockchain consisting of the block header and body fields, with each header containing Merkle root, the hash function of the previous block, data, and timestamp. The body contains transaction data connected to chains of blocks. Timestamp ensures the time created, while Merkle root verifies transaction integrity against the malicious configuration of previous block transactions. Changed blocks cause changes in hash functions, forming a fork, exposing the location of changed blocks, providing security against adverse, risks, and fulfilling the privacy demands of IoT. The EISO framework implements blockchain at edge cloudlets in the system to preserve privacy.

To demonstrate the effectiveness of the presented solution, this section provides a use case scenario of the IoT paradigm using an intelligent transportation system (ITS) where smart vehicles interact to share critical information and implement an autonomous edge-enabled IoT services provisioning scenario. The blockchain has been implemented at the edge cloudlets where the resources in the cloudlets are allocated to the IoT devices using the secure blockchain technique. The algorithm can fulfill the heterogeneous privacy demands of the ITS infrastructure. The system consists of three planes, including the infrastructure plane with resources such as edge cloudlets, IoT infrastructure, and forwarding devices like routers, switches, and gateways to guide network traffic. Blockchain is integrated at the edge cloudlets due to sufficient resources for handling transactions and challenging central cloud services with lower latency. Edge cloudlets support offloading services to IoT infrastructure with each cloudlet containing a block of overall blockchain. The working principle and overall structure of the solution have been explained.

Architecture

The algorithm uses the blockchain for secure computation offloading of latency-sensitive data generated by the IoT infrastructure and it employs private blockchain where mining is performed by resource-efficient edge servers. The architecture consists of an application plane, a control plane, and an infrastructure plane. The application plane offers a programmable platform for custom privacy-aware application development, whereas the control plane implements network policies and manages traffic forwarding requirements. The infrastructure plan includes physical hardware and comprises SDN switches and a cloud services layer that offer offloading services to perform compute-intensive tasks. The solution extends the EISO framework using an SDN-managed paradigm and implements the blockchain-based privacy enforcement solution.

The control plane implements network policies and manages traffic forwarding requirements. The controller is designed to serve a maximum number of requests without compromising QoS using virtual instances created using hypervisors. Multiple physical controllers are used to avoid a single point of failure. The control plane installs forwarding rules on the data plane devices using southbound RESTful API and acts as an intelligent platform with a global view of the network. The control plane monitors the status of forwarding devices at the data plane and implements the policies for accessing appropriate cloudlets and blockchain-based privacy enforcement. With this architecture, localized services for the IoT devices are provided, while the central controller handles all the remote controller instances.

The infrastructure plan includes hardware such as OpenFlow switches, routers, gateways, Road Side Unit (RSU)s, base stations, edge cloudlets, and end-IoT devices. For example, vehicles use a P2P network while other nodes use IEEE 802.11 to communicate. All nodes connect to the global controller using a long-range communication protocol such as Long-term Evolution Advanced (LTE-A). The infrastructure plane incorporates edge cloudlets to offer data computation and offloading capabilities to resource-constrained IoT infrastructures, enhancing privacy and integration in the IoT-edge paradigm. The infrastructure plane has two layers: SDN switches that forward traffic according to the controller's requirements and a cloud services layer offering offloading services for compute-intensive tasks.

Edge cloudlets act as RSUs authenticating vehicles in the ITS and creating, appending, and deleting blocks according to the proposed rules. Cloudlets act as authentication nodes that decide whether to allow or block incoming requests. The infrastructure plane communicates with edge cloudlets and IoT devices for privacy enforcement.

Authentication Mechanism

The blockchain is a distributed database that maintains a growing list of interconnected blocks, creating an independent ledger. Cloudlet blocks are identified by public keys, maintaining IoT-generated data privacy. Transactions in IoT use broadcasted public keys verified by edge cloudlets, comparing them with transaction generator signatures. Smart vehicles request secure transactions from edge cloudlets using decentralized blockchains, working together based on trust. Trusted nodes act as data trustees to keep data safe and block malicious traffic from adversaries.

In the above example, smart vehicles request V2V and V2I services from the ITS. Edge cloudlets offer offloading and data transfer capabilities to the cloud. Local controllers manage cloudlets and blockchain is implemented on every cloudlet controlled by localized controllers. The internal blockchain coin system links smart vehicles to edge cloudlet resources, ensuring seamless and secure service orchestration and provisioning. Mining is performed at the edge cloudlets with more available resources compared to resource-constrained IoT devices.

1.2.2.3 Working Principle of the Proposed Technique

This section presents the working principle of the algorithm for the above vehicle example. Smart vehicles request services from edge cloudlets, which are governed by controllers to offload tasks and ensure efficient data transfer. The blockchain is implemented on every edge cloudlet in the modified EISO framework supported by SDN for added security against information spoofing and data theft attacks. Traffic flow is initiated from smart vehicles, directed toward SDN switches, and subsequently forwarded to the controller. The controller is responsible for installing new flow rules that govern the handling of packets within the network. Blockchain clients perform computations on the edge cloudlets and identify transactions using the public key compared to the corresponding private key maintained by authenticate nodes. Edge cloudlets act as authenticate nodes, ensuring authenticity based on registration IDs and acting as blocks in the blockchain. They also act as data trustees that save data and protect against malicious traffic from adversaries.

Each cloudlet is managed by a distributed controller, which in turn is managed by the central controller to ensure efficient resource allocation. The incorporation of blockchain at the edge cloudlet layer provides privacy-enforcement mechanisms for secure service provisioning to resource-constrained devices. The request and response mechanism has also been discussed, which ensures the privacy of the entities while offering required services to the IoT devices.

In the solution, a secure transaction system for V2V and V2I communication is established through a request–response workflow. Smart vehicles request resources from edge cloudlets, and the request is processed by the blockchain to ensure data integrity using the public key. The private key associated with the public key is verified by the edge cloudlet, and the transaction is conducted if they match. After verification, the results are transferred to the requesting vehicle. Similarly, the response is authenticated by the blockchain before being delivered to the requesting station. The authentication mechanism ensures the privacy of the overall infrastructure, with every component working seamlessly together. The blockchain-based solution can provide detailed authentication mechanisms, which will be discussed further.

To ensure the privacy of the entire infrastructure, every component collaborates seamlessly. To maintain the privacy of the system, it is necessary to verify the identity and validity of a vehicle, as well as establish a method for revoking public keys. By utilizing blockchain technology, the need for third-party validation to enforce privacy can be eliminated, resulting in a more reliable and dependable system without single points of failure. To achieve this, information regarding vehicles' access or revocation from the network is transmitted to permission blockchain nodes known as authenticate nodes. Authentication of vehicles occurs at edge cloudlets, which are widely available throughout the ITS. In addition, a central controller analyzes vehicle credentials and coordinates with distributed localized controllers to allow access to the network. Once authenticated, the vehicle's public and private keys are stored where authenticate nodes are authorized to permit or reject further transactions. If a vehicle behaves unpredictably, the controller sends a notification to the blockchain where authenticate nodes decide whether to accept or reject according to pre-set rules. The admission/revocation information stored in the blockchain is utilized by vehicles to authenticate each other with minimal overhead. Each vehicle joining the network generates a public and private key; the public key is used for communication and maintaining integrity while ascertaining membership in the blockchain. Authenticate nodes issue validation certificates to legal vehicles, then deliver them to the blockchain via registration transaction.

1.3 Introduction to Business Process Management

1.3.1 Business Process Management (BPM)

Business process management (BPM) refers to a management discipline that focuses on improving business processes to enhance organizational efficiency and agility. BPM involves analyzing, modeling, optimizing, automating, and monitoring business processes, with the ultimate goal of creating value for customers and stakeholders.

BPM can be applied to various types of processes, such as operational processes, customer-facing processes, administrative processes, and strategic processes. BPM can also be used in different industries, such as manufacturing, healthcare, finance, and government. BPM is often supported by software tools and technologies, such as BPM suites, workflow systems, process mining tools, and analytics platforms. These tools provide capabilities for designing, executing, monitoring, and improving business processes throughout their lifecycle.

1.3.2 The Benefits and Drivers

BPM matters because it can help organizations achieve a wide range of benefits, such as improving efficiency and productivity by streamlining and automating business processes. Organizations can reduce cycle times, eliminate bottlenecks, and improve resource utilization, enhancing quality and consistency by standardizing and controlling business processes. Organizations can improve quality, consistency, and compliance with regulations and standards, increasing flexibility and agility by enabling rapid change and adaptation of business processes. Organizations can respond quickly to market demands, customer needs, and new opportunities. Moreover, by utilizing BPM, organizations can achieve enhanced customer experiences by aligning their business processes with customer needs and expectations. This customer-centric approach enables organizations to improve customer satisfaction, loyalty, and retention. By streamlining and optimizing processes to deliver exceptional customer experiences, businesses can strengthen their relationships with customers and drive long-term success.

The drivers for adopting BPM vary across organizations, but some common ones are cost reduction, competitive advantage, compliance, and risk management. BPM can help organizations reduce costs by eliminating waste, inefficiencies, and redundancies in business processes. It can offer a competitive advantage by enabling organizations to differentiate themselves from their competitors through superior customer service, faster time-to-market, and better quality. Moreover, BPM can help organizations comply with legal and regulatory requirements, as well as mitigate risks associated with business operations.

1.3.3 The Process Lifecycle and Methodologies

BPM involves a continuous cycle of activities spanning the business process lifecycle. These activities include process discovery, process modeling, process analysis, process redesign, process implementation, process monitoring, and process optimization.

Process discovery is identifying existing business processes and understanding their characteristics, stakeholders, and performance metrics. Process modeling aims to create abstract representations of business processes using graphical notations, such as business process modeling notation (BPMN), Unified Modeling Language (UML), or Event-driven Process Chain (EPC). Process analysis evaluates the effectiveness, efficiency, and feasibility of business processes, and identifies areas for improvement. Process redesign is modifying or reengineering business processes to optimize their performance and align them with organizational objectives. Process implementation tries to build or modify information systems and IT infrastructure to support the execution of business processes. Process monitoring tracks the performance of business processes in real-time, and detects issues, exceptions, and opportunities. Process optimization continuously improves the design and execution of business processes, based on feedback and data analysis.

To perform these activities effectively, organizations can use various BPM methodologies, such as Six Sigma, Lean, Agile, or Design Thinking. These methodologies provide frameworks, techniques, and tools for managing process improvement projects, engaging stakeholders, and achieving results.

1.3.4 Challenges and Best Practices in BPM

Despite its potential benefits, BPM faces several challenges that organizations need to overcome. Some common challenges are resistance to change, technical complexity, governance and ownership, data quality and availability, etc.

Resistance to change means most BPM initiatives require changes in organizational culture, behavior, and mindset, which may face resistance from employees, managers, or other stakeholders. BPM often involves integrating multiple information systems, data sources, and technologies to support business processes, which can be technically challenging and costly. Also, BPM requires clear roles, responsibilities, and accountability for process improvement activities, which may be difficult to define and communicate across organizational boundaries. It depends heavily on accurate, timely, and relevant data, but many organizations lack adequate data governance, management, or analytics capabilities.

To address these challenges, organizations can adopt best practices aligned with their goals, cultures, and contexts. Establishing a process-centric culture involves promoting a mindset that values continuous improvement, innovation, collaboration, and customer focus. Defining clear process ownership involves assigning responsibility for specific processes to individuals or teams, and empowering them to make decisions, solve problems, and drive change. Leveraging technology and data involves selecting appropriate tools and platforms that support process modeling, analysis, execution, and monitoring, as well as ensuring data accuracy, completeness, and security. Engaging stakeholders involves all relevant stakeholders in process improvement initiatives, such as customers, suppliers, partners, employees, and managers, and communicating transparently and effectively with them.

1.4 Summary

The first section presents the EISO framework for optimal resource allocation in IoT services. Our approach utilizes a coordinated effort between mist computing, edge cloudlets, and central cloud to provide QoS-aware resource provisioning. The framework hypothesizes that latency, delay, and service delivery rate are critical factors influencing QoS-aware service provisioning. Therefore, this book devises an analytical framework that formulates resource allocation as an optimization problem that maximizes network utility.

The framework optimally allocates resources for delivering efficient services to requesting stations. This framework can greatly reduce energy consumption, and service latency, and enhances service delivery rate in edge-enabled IoT paradigms. Despite these significant benefits, enormous challenges still exist during service orchestration, including heterogeneity in communication infrastructure and platform-specific issues. As such, application developers and businesses need to consider these challenges when developing new applications for customized and user-based service orchestration.

Then the next section presents a network-level security solution designed for the application layer of the EISO framework, to address the lack of security algorithms and solutions protecting IoT devices with limited computational resources. This book utilizes the EISO framework for the orchestration of IoT services, investigates the impact of DDoS attacks on the data plane, and introduces a defense mechanism for data plane security in the EISO framework. Unlike previous techniques involving complex network measurements using specialized hardware or software agents, the new model provides defense against DDoS attacks without requiring additional hardware for traffic measurement. Thus, it ensures comprehensive security without imposing extra overhead on the network.

While security-enforcement techniques ensure the reliability of the EISO framework, privacy concerns related to data transactions in IoT service delivery that traverse multiple network infrastructure poses challenges. Furthermore, data processing on edge devices can lead to privacy leakage due to multi-device interactions. Therefore, the framework addresses privacy issues during IoT service orchestration and introduces a distributed authentication strategy as a solution for the edge-enabled IoT service orchestration framework presented in this study.

Also, this section introduces a privacy-preserving technique for the EISO framework by deploying blockchain at the edge layer. The introduced algorithm leverages the decentralized security mechanism of blockchain to establish a privacy-preserving ledger for edge-enabled IoT services. To enforce trust in IoT devices, a decentralized trust-management paradigm is implemented on the edge nodes using previous interactions between IoT devices and the edge cloudlets. Trusted edge nodes utilize an authentication algorithm for privacy-preserving service provisioning to users. Traditional privacy-preserving techniques such as Public Key Infrastructure (PKI) are prone to single-point-of-failure and complex computations, thus, the presented algorithm utilizes the decentralized blockchain mechanism for enhanced privacy at the edge-enabled IoT. The utilization of edge cloudlets enables efficient end-to-end IoT service orchestration while maintaining fault tolerance, reliability, and trustworthiness. Blockchains act as authenticators to permit transactions, enabling efficient privacy preservation in the edge-enabled IoT paradigm. By employing the decentralized blockchain mechanism at the edge layer, our solution offers an efficient solution for privacy-preserving IoT services.

The second section highlights the significance of the EISO framework for end-to-end IoT service orchestration, and its key technologies focused on application development, security, and privacy. This research contributes to the existing body of knowledge by presenting a comprehensive solution to efficiently manage edge-enabled IoT services. Moreover, the presented solution ensures the security of services while preserving the privacy of users in a heterogeneous IoT environment. Therefore, this study is a crucial step toward achieving efficient end-to-end IoT service orchestration while maintaining security and privacy for all stakeholders involved in the delivery of IoT services. In conclusion, the EISO framework offers a reliable and trustworthy solution for conducting IoT services in an integrated and coordinated manner for end-to-end IoT service orchestration.

The last section introduces business process management (BPM); BPM is a discipline that has gained increasing attention from organizations of all sizes and types, as they seek to improve their performance, competitiveness, and customer experience. BPM involves a set of activities and methodologies that help organizations analyze, model, optimize, automate, and monitor their business processes, to create value for customers and stakeholders. To realize the benefits of BPM, organizations need to overcome various challenges and adopt best practices that are tailored to their goals, cultures, and contexts.

References

1. 李进(Wajid Rafique) (2020) 支持边缘计算的物联网服务协同框架及其关键技术研究[D] 南京大学. https://doi.org/10.27235/d.cnki.gnjiu.2020.002576
2. Peng K, Huang H, Wan S, Leung VC (2020) End-edge-cloud collaborative computation offloading for multiple mobile users in heterogeneous edge-server environment. Wirel Netw:1–12
3. Samanta A, Jiao A, Mühlhäuser M, Wang L (2019) Incentivizing microservices for online resource sharing in edge clouds. In: Proceeding of the IEEE 39th International Conference on Distributed Computing Systems (ICDCS), Dallas, TX, pp 420–430
4. Xu X, Sun W, Vivekananda GN, Shankar A (2020) Achieving concurrency in cloud-orchestrated internet of things for resource sharing through multiple concurrent access. Comput Intell:1–16
5. Martinez I, Jarray A, Hafid AS (2020) Scalable design and dimensioning of fog-computing infrastructure to support latency sensitive IoT applications. IEEE Internet Things J 7(6): 5504–5520
6. Arkian HR, Diyanat A, Pourkhalili A (2017) MIST: fog-based data analytics scheme with cost-efficient resource provisioning for IoT crowdsensing applications. J Netw Comput Appl 82: 152–165

7. Sharma S, Saini H (2020) Fog-assisted task allocation and secure deduplication using 2FBO2 and MoWo in cluster-based industrial IoT (IIoT). Comput Commun 152:187–199
8. Mishra S, Sahoo MN, Bakshi S, Rodrigues JJ (2020) Dynamic resource allocation in fog-cloud hybrid systems using multi-criteria AHP techniques. IEEE Internet Things J 7(9):8993–9000
9. Zhang T, Jin J, Zheng X, Yang Y (2019) Rate-adaptive fog service platform for heterogeneous IoT applications. IEEE Internet Things J 7:176–188
10. Zhou Z, Yu S, Chen W, Chen X (2020) CE-IoT: cost-effective cloud-edge resource provisioning for heterogeneous IoT applications. IEEE Internet Things J 7(9):8600–8614
11. Asif-Ur-Rahman M, Afsana F, Mahmud M, Kaiser MS, Ahmed MR, Kaiwartya O et al (2018) Toward a heterogeneous mist, fog, and cloud-based framework for the internet of healthcare things. IEEE Internet Things J 6:4049–4062
12. Bui K-HN, Jung JJ (2019) Computational negotiation-based edge analytics for smart objects. Inf Sci 480:222–236
13. Ferdowsi A, Challita U, Saad W (2019) Deep learning for reliable mobile edge analytics in intelligent transportation systems: an overview. IEEE Veh Technol Mag 14:62–70
14. Yang P, Lyu F, Wu W, Zhang N, Yu L, Shen XS (2019) Edge coordinated query configuration for low-latency and accurate video analytics. IEEE Trans Industr Inform 16:4855–4864
15. Anagnostopoulos (2020) Edge-centric inferential modeling and analytics. J Netw Comput Appl 164:1–21
16. Garg S, Singh A, Kaur K, Aujla GS, Batra S, Kumar N et al (2019) Edge computing-based security framework for big data analytics in VANETs. IEEE Netw 33:72–81
17. Yassine S, Singh MS, Hossain MS, Muhammad G (2019) IoT big data analytics for smart homes with fog and cloud computing. Futur Gener Comput Syst 91:563–573
18. Lu Z, Wang N, Wu J, Qiu M (2018) IoTDeM: an IoT big data-oriented MapReduce performance prediction extended model in multiple edge clouds. J Parallel Distributed Comput 118: 316–327
19. Wan S, Lu J, Fan P, Letaief KB (2020) Towards big data processing in IoT: path planning and resource management of UAV base stations in mobile-edge computing system. IEEE Internet Things J 7(7):5995–6009
20. Xu C, Wang K, Li P, Guo S, Luo J, Ye B et al (2018) Making big data open in edges: a resource-efficient blockchain-based approach. IEEE Trans Parallel Distributed Syst 30:870–882
21. Gennaro R, Gentry C, Parno B (2010) Non-interactive verifiable computing: outsourcing computation to untrusted workers. In: Proceedings of the Springer Annual Cryptology Conference , Santa Barbara, CA, USA, pp 465–482
22. Pan J, Wang J, Hester A, AlQerm I, Liu Y, Zhao Y (2018) EdgeChain: an edge-IoT framework and prototype based on blockchain and smart contracts. IEEE Internet Things J 6:4719–4732
23. Makhdoom I, Zhou M, Abolhasan M, Lipman J, Ni W (2020) PrivySharing: a blockchain-based framework for privacy-preserving and secure data sharing in smart cities. Comput Secur 88: 101653
24. Ma M, He D, Wang H, Kumar N, Choo K-KR (2019) An efficient and provably secure authenticated key agreement protocol for fog-based vehicular ad-hoc networks. IEEE Internet Things J 6:8065–8075
25. Yang J, Weng N, Cheng J, Ni XL, Shen X (2019) Deqos attack: degrading quality of service in vanets and its mitigation. IEEE Trans Veh Technol 68:4834–4845

Chapter 2
Architecture of Industrial Internet-Centric BPM

Abstract Industrial Internet-centric BPM (business process management) refers to the application of business process management principles and technologies in the context of the Industrial Internet. It focuses on optimizing and automating business processes within industrial environments by leveraging the power of data, connectivity, and advanced analytics. This approach combines traditional BPM concepts with Industrial Internet of Things (IIoT) technologies to enable real-time monitoring, analysis, and control of complex industrial processes. It aims to improve operational efficiency, productivity, and decision-making by integrating business systems with operational systems.

Overall, Industrial Internet-centric BPM empowers organizations to transform their industrial processes by combining BPM methodologies with the capabilities of the Industrial Internet, ultimately driving operational excellence and competitive advantage. In the following sections, this book first introduces mobile Industrial Internet-centric BPM, and the next section discusses resource deployment in BPM. The last section considers how edge computing can be integrated into existing frameworks to allow for more efficient use of resources. The architecture is designed to reduce latency and improve resource utilization, ensuring optimal performance of the business process scheduling problem.

Industrial Internet-centric BPM (business process management) refers to the application of business process management principles and technologies in the context of the Industrial Internet. It focuses on optimizing and automating business processes within industrial environments by leveraging the power of data, connectivity, and advanced analytics. This approach combines traditional BPM concepts with Industrial Internet of Things (IIoT) technologies to enable real-time monitoring, analysis, and control of complex industrial processes. It aims to improve operational efficiency, productivity, and decision-making by integrating business systems with operational systems.

Industrial Internet-centric BPM utilizes sensors, automation devices, and data streams to capture and analyze relevant process information. It allows for better visibility into the end-to-end operations, identifies bottlenecks or inefficiencies, and enables proactive decision-making to optimize workflow and resource allocation. By

W. Dou et al., *Intelligent Industrial Internet Systems*,
https://doi.org/10.1007/978-981-99-5732-3_2

leveraging advanced analytics techniques such as machine learning and predictive modeling, Industrial Internet-centric BPM can provide insights for process optimization, predictive maintenance, quality control, and risk management. It enables organizations to achieve higher levels of agility, flexibility, and responsiveness in their industrial operations.

Overall, Industrial Internet-centric BPM empowers organizations to transform their industrial processes by combining BPM methodologies with the capabilities of the Industrial Internet, ultimately driving operational excellence and competitive advantage. In the following sections, this book first introduces mobile Industrial Internet-centric BPM, and the next section discusses resource deployment in BPM. The last section considers how edge computing can be integrated into existing frameworks to allow for more efficient use of resources. The architecture is designed to reduce latency and improve resource utilization, ensuring optimal performance of the business process scheduling problem.

2.1 Mobile Industrial Internet-Centric BPM

This section introduces an intelligent architecture that prioritizes latency guarantee and storage capacity occupation during communication and transferring operations in a location-aware environment. In summary, this section emphasizes the challenges posed by the limited capacity of edge servers, which creates a trade-off between mobile devices, edge servers, and cloud data centers to provide efficient low-latency computing. It also points out the key issues not addressed in existing research on server selection and proposes an intelligent architecture that focuses on latency guarantee and storage capacity occupation during communication and transferring operations. This chapter is significant because it provides insights into location-aware computing environments and addresses multiuser server selection problems through a distributed service transfer algorithm.

2.1.1 Introduction

The Industrial Internet of Things (IIoT) improves production efficiency, reduces errors, and enhances security by embedding intelligent devices into the life cycle of industrial products. IIoT business processes are frequently used in logistics management, manufacturing process optimization, safety monitoring, etc. However, the increasing amount of mobile data generated by IIoT devices presents significant challenges to data management. Cloud-edge collaboration provides a solution for these challenges by combining cloud computing and edge computing paradigms. A good strategy for data placement is essential in improving access efficiency and reducing costs in the overall running time of the IIoT business process.

With the rise of mobile applications, terminal devices cannot often handle complex business task computing, leading to offloading tasks to the cloud or data center. However, for IIoT applications like intelligent self-driving vehicles and unmanned aircraft systems that require low latency, edge computing provides a solution by utilizing distributed edge servers. Nevertheless, the limited computing and storage capacity of an edge server necessitates a trade-off between mobile devices, edge servers, and the cloud data center for efficient low-latency computing. Server selection has been studied extensively in various fields like network latency and load-balancing, but four key issues remain unaddressed. First, the mobility of devices and communication with servers in a location-aware environment must consider storage capacity occupation during communication. Second, service providers seek the best-performing edge server along users' paths, but edge servers' limited capacity necessitates a cost-effective transfer strategy. Third, some works only focus on a single user or users with a single task, ignoring burden distribution among different users. Fourth, using centralized strategies or protocols cannot address the multiuser server selection problem, requiring a distributed service transfer algorithm.

Mobile Industrial Internet-centric BPM refers to the integration of mobile device technologies with the Industrial Internet of Things (IIoT) and business process modeling (BPM) in industrial settings. This integration enables the use of mobile devices such as smartphones and tablets to access and manage IIoT data in real time, thus providing a more agile, flexible, and responsive approach to industrial process management. By combining mobile technology with IIoT and BPM, it is possible to enhance collaboration among stakeholders, improve the visibility of industrial processes, and drive operational efficiency. Mobile Industrial Internet-centric BPM can be applied in diverse areas such as manufacturing, logistics, and supply chain management, with the common goal of optimizing production processes, reducing errors, and enhancing security. Although this concept is relatively new, it has enormous potential to revolutionize the way industries function and become a game changer in Industrial Internet applications.

2.1.2 *Server Selection Example*

2.1.2.1 Formulation of Server Selection Problem

A typical location-aware environment scenario involves base stations, edge servers, and mobile devices. Figure 2.1 demonstrates the interaction between these components. The mobile device receives services via the base station from an edge server. In a location-aware environment, the base station can sense a user's location as they move from one coverage area to another. The base station also senses the direction of movement and data transmission rate for the device, which is sent to the serving edge server for the task.

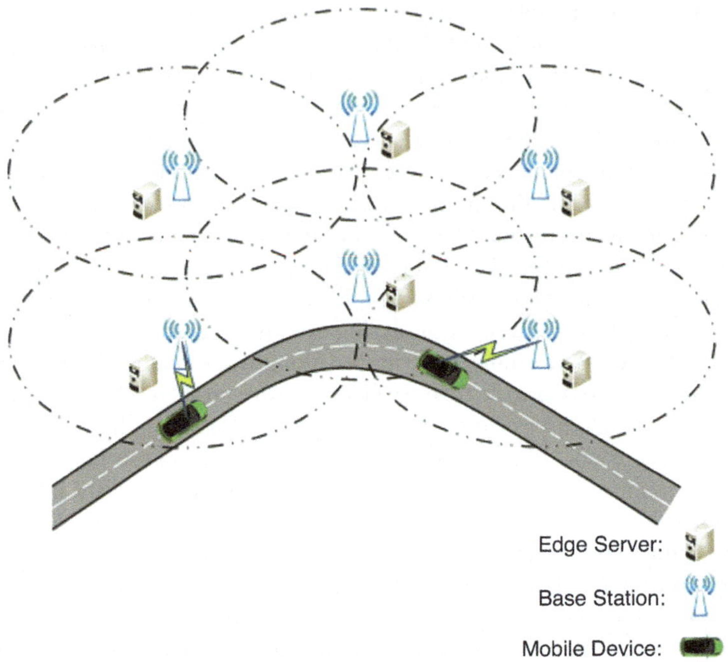

Fig. 2.1 A scenario of a location-aware environment

When a user task changes location, the edge server can determine whether the task needs to be migrated based on the information provided by the base station. A candidate server for migration is also selected based on perceived information. The choice of server depends on several factors such as network latency, available resources, and server load. This process guarantees that when a user moves into a different coverage area, the most suitable server can provide uninterrupted service.

The location-aware environment considers a dense cellular network layout for base stations as shown in Fig. 2.1. In this scenario, the user-level server selection problem is addressed, which requires the base station to sense the user's direction of movement, speed, and data transfer rate. Server selection modeling is performed for this scenario in line with this assumption. The server that provides services must meet specific performance criteria such as low latency, high availability, and the least possible cost. Overall, the location-aware environment concept improves efficiency by providing seamless service delivery across a network of edge servers.

Edge computing allows mobile devices to offload computing tasks to edge servers and cloud centers [1]. At present, edge computing is widely associated with the Internet of Vehicles [2, 3], the Industrial Internet of Things [4], health care [5, 6], and other industries [7–9]. The topic of server selection and service transfer has been intensively explored through cloud computing during the last decade [10–12]. One concern is that cloud computing service transfer is not very sensitive to delays in terms of transfer technology. Edge computing tasks are more susceptible to delays and cloud computing's service transfer technology cannot fulfill task delay requirements. Some new transfer techniques [13–15] are found to solve the above problems.

2.1.2.2 Architecture

This section outlines the goals and presents formulas for a mobile environment [16], but notes that the EC scenarios are subject to frequent changes, making it challenging to apply real-world settings. The cost model is estimated without considering edge server load, and in high-traffic situations, directly applying ideal conditions may result in server overload. However, candidate server priority obtained from ideal conditions can still be used to design a suitable service transfer method by solving the multiuser server selection problem using scheduling strategies.

When a task is executed in the cloud, the algorithm can migrate tasks to edge servers without historical data. On an edge server, historical user task data are known and can be used to determine migration timing and candidate server priority. This process is shown in Fig. 2.2.

The algorithm is primarily made up of five pieces that work together. S_{source} represents the user's source edge server, whereas $S_{candidate}$ represents the candidate edge server to which u_j might transfer. t_j executes the following steps:

Step 1: S_{source} identifies the location information of t_j.

Step 2: S_{source} collects all tasks' location information. According to the task's location information, determine whether the task needs to be transferred. If the task reaches the optimal transfer location, S_{source} calculates the priority of the task's candidate servers. Then, S_{source} sends the request to the highest priority server in the candidate server list. In a time slot, S_{source} will only send the request once for a task. If the task is not in the optimal transfer position, S_{source} will continue identifying the task location information.

Step 3: $S_{candidate}$ collects the requests in a time slot. Based on the number of available resources to $S_{candidate}$, $S_{candidate}$ decides the number of resources to allocate to S_{source} and returns the pre-allocated amount of resources to S_{source}. For a limited time, $S_{candidate}$ locks resources.

Step 4: If the pre-allocated resource is sufficient, S_{source} will decide which tasks to transfer according to the task priority. Otherwise, the tasks that fail to transfer will remove the highest priority server from its candidate server list. Then, these tasks will participate in the transfer in the next time slot in Step 2. The remaining resources of the pre-allocated resource will be unlocked at the end of the time slot.

This architecture focuses on the mobile IIoT environment. In this environment, the architecture can deliver edge service to the majority of users at the lowest system cost. Locking actions during resource allocation can also prevent restricted resources from being assigned to users.

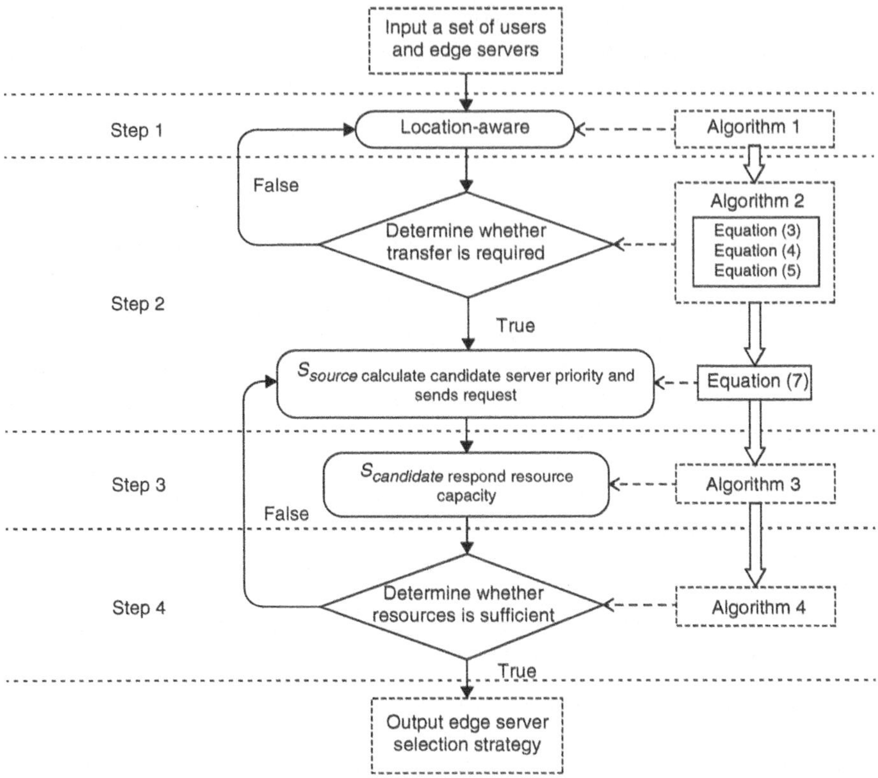

Fig. 2.2 The architecture for mobile business task computing

2.2 Resources Deployment in BPM

In today's fast-paced business environment, companies are constantly seeking ways to improve their operational efficiency and reduce costs while delivering high-quality products and services to their customers. Resources deployment plays a crucial role in achieving these goals. Efficient deployment of resources helps ensure that tasks are completed on time, costs are kept under control, and customer satisfaction is kept high. The purpose of this section is to provide an introduction to resource deployment in BPM. This section starts by introducing key concepts and models related to BPM, followed by an overview of resource deployment and its importance in BPM. It then shows the architecture of resource deployment in BPM.

Business process management (BPM) is a popular technique that allows organizations to optimize their business operations by defining, modeling, executing, and monitoring processes. One of the essential components of BPM is resource deployment, which deals with the allocation of resources such as people, equipment, and facilities to complete tasks within a process. The use of distributed computing systems has increased due to the exponential growth of data in the age of information. One of the key metrics for assessing their effectiveness is job completion time (JCT). Data skew is a common problem that affects the performance of these systems, leading to the need to reduce JCT through reasonable scheduling. This section introduces the architecture of resource deployment in BPM to reduce JCT by handling data skew during the reducing phase of processing.

2.2.1 Introduction

BPM is a holistic approach to managing business processes that aim to improve operational efficiency, reduce costs, and enhance customer satisfaction. As mentioned in Chap. 1, the key concepts of BPM are process discovery, process modeling, process analysis, process redesign, process implementation, process monitoring, and process optimization.

Distributed computing systems like Hadoop, Spark, and Hive are increasingly used due to the exponential increase in data. They have high reliability, high scalability, high fault tolerance, and efficiency, and can support processing very large datasets across server clusters. Job completion time (JCT) is a major metric used to evaluate these systems, and in this section, the focus of the architecture is on minimizing JCT. Data skew, which can compromise the performance of distributed computing systems, is also addressed. However, it can be found that the default scheduling mechanism for reducing tasks in Hadoop does not take the data skew into consideration, and simply schedules all reduce tasks in a first-in-first-out (FIFO) manner.

Besides, some works are addressed on achieving high network utilization and reducing communication time [17–19]. In directed acyclic graph, scheduling heterogeneous DAG can not only improve the throughput of cluster but also reduce JCT [20]. Reinforcement learning [21] is also used to learn sophisticated scheduling policies automatically to reduce JCT. Optimizing memory management policy in Java virtual machine [22, 23] can reduce garbage collection time, thus reducing JCT. Merging fragmented intermediate files [24] can convert small, random disk IO requests to large, sequential ones, thus reducing completion time.

The largest-first mechanism [25] is presented as a helpful scheduling approach to improve performance by scheduling reduce tasks with longer execution time first. To illustrate issues related to scheduling reduce tasks with skewed input data, this section uses Fig. 2.3 which shows each slice representing a reduce task, with its label indicating the duration. The aim is to overlap large reduce tasks with smaller ones as much as possible to reduce job completion time (JCT) effectively. The example provided assumes three reduce tasks with each task taking 3 T, 2 T, and 1 T respectively, where T represents the basic time unit. All educe tasks executing the same code for the same job will be queued in Hadoop waiting for slots to be released. In this particular example, three slots are released at 1 T, 2 T, and 3 T respectively to reduce tasks to run.

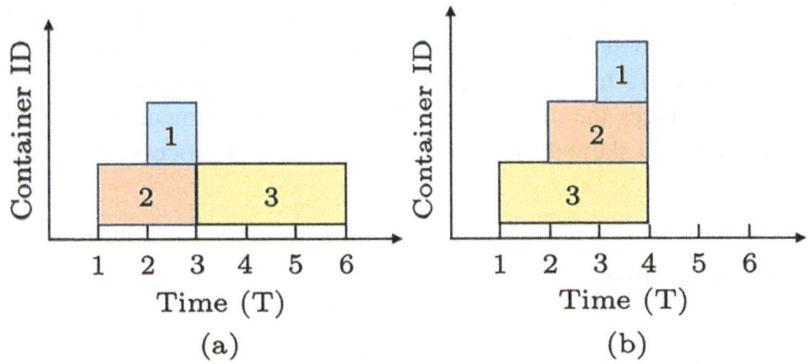

Fig. 2.3 Example of reduce task scheduling. (**a**) FIFO. (**b**) Largest-first scheduling

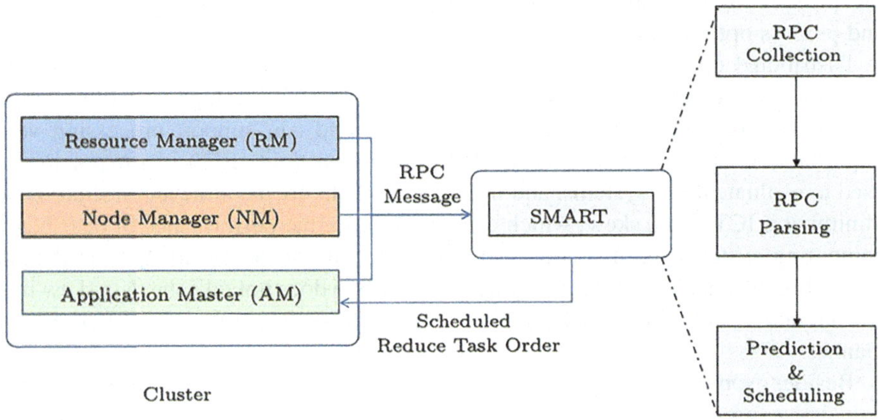

Fig. 2.4 Overall architecture

Figure 2.3 displays the task execution details for both FIFO and largest-first scheduling approaches, with the former scheduling order being {T2, T1, T3}, and the latter ordering being {T3, T2, T1}. The result demonstrates that the largest-first approach takes 3 T executing time while FIFO takes 5 T, a reduction of 40% by using largest-first scheduling. So that this section introduces an architecture that utilizes largest-first scheduling to effectively improve the overall performance of distributed computing systems during the reducing phase of MapReduce jobs.

2.2.2 Architecture

As shown in Fig. 2.4, the cluster is mainly composed of the resource manager (RM), node manager (NM), application master (AM), and containers. The fundamental idea is to split up the functionalities of resource management and job scheduling/

monitoring into separate modules. The overall architecture is positioned on a separate node outside the data processing cluster. It continuously monitors Remote Procedure Call (RPC) messages in the job to gather running task states, collects output sizes of completed map tasks, and predicts remaining reduce task sizes. Notably, this process relies solely on RPC message collection and parsing without requiring any program modifications, and the linear algorithms for prediction and scheduling ensure that the architecture operates efficiently. If there are too many jobs competing for the system, the residence node can be extended to enhance performance. To obtain intermediate data output size from map tasks, the architecture must acquire the total number of map tasks and individual tasks' running states and positions. In order to minimize revisions to the existing system, an RPC collection module was developed to access all RPC messages, while a parsing module obtained the necessary information for each map task. Furthermore, a prediction module was created to predict reduce task sizes based on the size of the intermediate results produced by 5% of all map tasks.

For prediction and scheduling, in Hadoop, implementing a reduce task size-based scheduling mechanism is challenging because the sizes of reduce tasks often remain unknown in advance. However, this architecture utilizes completed map tasks to predict reduce task sizes and reschedule accordingly. It should be noted that it predicts relative size relationships instead of precise sizes for reduce tasks. The underlying concept is that the intermediate accumulated partitions' size relationship is likely to align with the reduce task size relationship. Then the architecture uses predicted size relationships to implement the largest-first scheduling mechanism instead of the default FIFO scheduling in Hadoop. The largest-first scheduling algorithm sorts jobs by their task sizes and assigns the job with the longest processing time to available containers first.

It predicts reduce task size based on completed map tasks and employs largest-first scheduling accordingly. This architecture is not specific to this system but can be applied to any data processing system with a Directed Acyclic Graph (DAG) pattern and limited processing units. MapReduce jobs in Hadoop are a specific kind of DAG task where inputs of reduce tasks are outputs of map tasks. Moreover, works on predicting the reduce task size more precisely by adding an offline training module is still worthy to research.

2.3 Business Process Scheduling Problem

The business process scheduling problem involves optimizing the execution of a set of tasks to achieve specific goals, such as minimizing the time required to complete all tasks or reducing resource usage. In this context, a task can represent any work activity that needs to be performed, and resources can refer to people, machines, software systems, or other operational elements. The goal of business process scheduling is to allocate resources efficiently to minimize cost and maximize productivity. This requires careful management of dependencies between tasks, resource constraints, and timing considerations. Effective scheduling is critical to ensuring that business processes are executed smoothly and that organizations can meet their objectives in a timely and cost-effective manner. This section provides an elaborate overview of the business process scheduling problem and presents a framework for its optimization. Specifically, it considers how edge computing can be integrated into existing frameworks to allow for more efficient use of resources. The architecture is designed to reduce latency and improve resource utilization, ensuring optimal performance of the business process scheduling problem.

2.3.1 Introduction

Business process scheduling problem is a challenging task in the field of operational research and management science. It has become increasingly significant as digital transformation and automation have taken over industries. Business process scheduling refers to the allocation of resources to tasks to optimize productivity, reduce costs, and meet deadlines. This is achieved by allocating resources such as manpower, machines, and software systems in the best possible way, to ensure the efficient execution of all activities that make up the business process.

The business process scheduling problem typically involves numerous interdependent tasks that require coordination and proper sequencing. Each task requires specific resources, such as processing time, manpower, and tools. Under this scenario, the objective is to allocate the available resources effectively to minimize the overall schedule length, cost, and risks associated with the project. Effective scheduling is essential in ensuring that the projects are executed smoothly and deadlines are met on time. Ineffective scheduling can result in increased costs due to the use of excessive resources or penalties imposed for late submission of deliverables.

Traditionally, business process scheduling problems rely on centralized computing systems to allocate resources. Several challenges arise when dealing with large-scale processes and dynamic resource allocation at the network edge. As a result, traditional centralized computing systems may not fulfill requirements for low-latency applications. For instance, real-time monitoring and decision-making systems require immediate feedback, which may be hindered by challenges of high network latency, security concerns, and limited bandwidth.

2.3.2 Architecture for Business Process Scheduling Problem

As described above, the optimization of the business process scheduling problem can be done with the following steps as shown in Fig. 2.5: task identification, resource allocation, task prioritization, schedule generation, monitoring and control, and optimization phase.

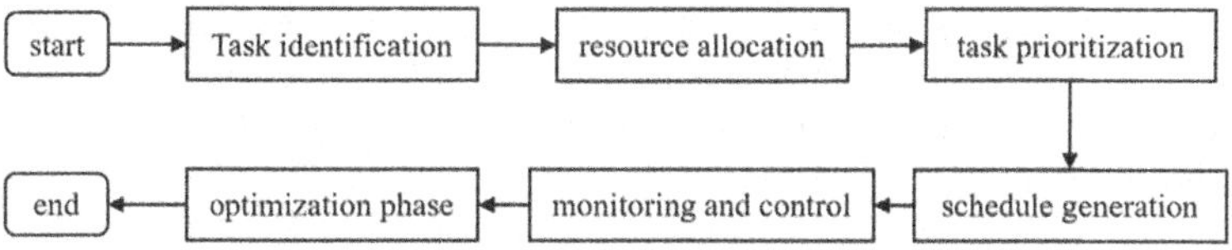

Fig. 2.5 Architecture of business process scheduling optimization

Task identification: The first step in business process scheduling is to identify all tasks involved. Typically, a business involves multiple tasks, and they are seldom run independently. Therefore, it is necessary to carefully consider the dependencies and sequence between each task during this stage. Before task identification, process analysis is conducted, and the overall business process is defined.

Resource allocation: This step involves identifying the available resources, and nowadays these resources include not only personnel, facilities, and tools but also software systems and technical equipment. Assigning necessary resources to each task can significantly improve efficiency. However, when allocating resources, it is important to avoid redundant use of resources and comply with restrictions on natural resource utilization. In addition, some companies may need to consider special business needs such as environmental protection and lean production principles.

Task prioritization: This step involves sorting tasks according to priority or urgency based on their dependencies and time costs. For example, the "first in, first out" principle for completing tasks might need to be followed. Also "Largest-first scheduling" introduced in Sect. 2.2 would produce better performance. Time costs should also factor in organizational content limits, resource constraints, and competition between other projects or tasks.

Schedule generation: Once the order of all tasks and available resources are determined, a schedule can be generated to optimize start and end dates and times for each task. When generating the schedule, most issues involve assigning tolerances for adjustments rather than determining planning slack, because other resources like personnel, materials, and equipment must be coordinated. Therefore, different scenarios need to be written up based on various coordination conditions when devising plans. This will include planning changes such as lost work hours and expanded resource consumption. Different goals in management plans may require adjustment based on other factors.

Monitoring and control: The monitoring and control phase is essential for the effective execution of task and time plans. During this process, important variables are monitored and data flow is tracked, and plans are updated and adapted promptly. In addition, the team often identifies new requirements and needs to use certain documents to track these requirements and other changes. All aspects related to quality and quantity should be recorded and re-arranged after planning changes.

Optimization: Optimization is the most critical part of this process, which can help organizations improve plan efficiency and form. To some extent, the optimization goal is a subjective choice, depending on how success is defined and considering the needs of all involved parties. Optimization strategies include but are not limited to rescheduling task spacing or urgency, reallocation of resources to reduce costs and maintain project scope, or fine-tuning schedules for greater efficiency and flexibility.

2.4 Summary

This chapter first introduces the mobile Industrial Internet-centric BMP and gives a specific example of how a server is selected in a location-aware environment scenario. Also, a detailed algorithm is presented to solve the server selection problem. Then this book introduces the resource deployment in BPM with an overall architecture that is not specific to this system but can be applied to any data processing system with a DAG pattern and limited processing units to reduce the job completion time. Finally, this chapter introduces a classical problem called the business process scheduling problem. The solution includes task identification, resource allocation, task prioritization, schedule generation, monitoring and control, and optimization. Solving business process scheduling problems requires thorough analysis and careful consideration at each stage, and management plans should be tailored to various experiences.

References

1. Wang T, Bhuiyan MZA, Wang G, Qi L, Jie W, Hayajneh T (2019) Preserving balance between privacy and data integrity in edge-assisted internet of things. IEEE Internet Things J 7(4): 2679–2689
2. Xu X, Fang Z, Qi L, Zhang X, He Q, Zhou X (2021) TripRes: traffic flow prediction driven resource reservation for multimedia IoV with edge computing. ACM Trans Multimed Comput Commun Appl (TOMM) 17(2):1–21
3. Xu X, Fang Z, Zhang J, He Q, Dongxiao Y, Qi L, Dou W (2021) Edge content caching with deep spatiotemporal residual network for IoV in Smart City. ACM Trans Sensor Netw (TOSN) 17(3):1–33
4. Zhao H, Liu Z, Tang J, Gao B, Zhang Y, Qian H, Huaqiang W (2022) Memristor-based signal processing for edge computing. Tsinghua Sci Technol 27(3):455–471
5. Xu X, Tian H, Zhang X, Qi L, He Q, Dou W (2022) DisCOV: distributed COVID-19 detection on x-ray images with edge-cloud collaboration. IEEE Trans Serv Comput 15(3):1206–1219
6. Wang X, Zhou Y, Zhao C (2022) Heart-rate analysis of healthy and insomnia groups with detrended fractal dimension feature in edge. Tsinghua Sci Technol 27(2):325–332
7. Wang F, Wang L, Li G, Wang Y, Lv C, Qi L (2021) Edge-cloud-enabled matrix factorization for diversified APIs recommendation in mashup creation. World Wide Web 24:1–21
8. Shen S, Huang L, Zhou H, Shui Y, Fan E, Cao Q (2018) Multistage signaling game-based optimal detection strategies for suppressing malware diffusion in fog-cloud-based IoT networks. IEEE Internet Things J 5(2):1043–1054

9. Shen Y, Shen S, Zongda W, Zhou H, Shui Y (2022) Signaling game-based availability assessment for edge computing-assisted IoT systems with malware dissemination. J Inf Secur Appl 66:103140
10. Chang H, Liu H, Leung Y-W, Chu X (2014) Minimum latency server selection for heterogeneous cloud services. IEEE Global Commun Conf 2014:2276–2282
11. Meiren T, Friedrich M, Schiller C (2021) Transfer of Service Research into Companies. Springer; International Conference on Applied Human Factors; Ergonomics, pp 400–405
12. Wendell P, Jiang JW, Freedman MJ, Rexford J (2010) Donar: Decentralized Server Selection for Cloud Services. In: Proceedings of the ACM SIGCOMM 2010 Conference, pp 231–242
13. Barbalace A, Karaoui ML, Wang W, Xing T, Olivier P, Ravindran B (2020) Edge Computing: The Case for Heterogeneous-Isa Container Migration. In: Proceedings of the 16th ACM SIGPLAN/SIGOPS International Conference on Virtual Execution Environments, pp 73–87
14. Qiu Y, Lung C-H, Ajila S, Srivastava P (2019) Experimental evaluation of LXC container migration for cloudlets using multipath TCP. Comput Netw 164:106900
15. Tang Z, Zhou X, Zhang F, Jia W, Zhao W (2018) Migration modeling and learning algorithms for containers in fog computing. IEEE Trans Serv Comput 12(5):712–725
16. Dou W, Liu B, Duan J, Dai F, Qi L, Xu X (2022) SeeMe: an intelligent edge server selection method for location-aware business task computing over IIoT. Software: Practice and Experience. https://doi.org/10.1002/spe.3179
17. Mosharaf C, Ion S (October 2012) Coflow: A networking abstraction for cluster applications. In: Proceeding of the 11th ACM Workshop on Hot Topics in Networks, pp 31–36
18. Chowdhury M, Zhong Y, Stoica I (2014) Efficient coflow scheduling with Varys. ACM SIGCOMM Comput Commun Rev 44(4):443–454
19. Chowdhury M, Stoica I (2015) Efficient coflow scheduling without prior knowledge. In: Proceeding of the 2015 ACM Conference on Special Interest Group on Data Communication, pp 393–406
20. Grandl R, Kandula S, Rao S, Akella A, Kulkarni J (November 2016) Graphene: Packing and dependency-aware scheduling for data-parallel clusters. In: Proceeding of the 12th USENIX Conference on Operating Systems Design and Implementation, pp 81–97
21. Mao H, Schwarzkopf M, Venkatakrishnan SB, Meng Z, Alizadeh M (2019) Learning scheduling algorithms for data processing clusters. In: Proceeding of the ACM Special Interest Group on Data Communication, pp 270–288
22. Nguyen K, Wang K, Bu Y, Fang L, Hu J, Xu G (2015) FACADE: A compiler and runtime for (almost) object-bounded big data applications. In: Proceeding of the 20th International Conference on Architectural Support for Programming Languages and Operating Systems, pp 675–690
23. Nguyen K, Fang L, Xu G, Demsky B, Lu S, Alamian S, Mutlu O (2016) Yak: A high-performance big-data-friendly garbage collector. In: Proceeding of the 12th USENIX Symposium on Operating Systems Design and Implementation, pp 349–365
24. Zhang H, Cho B, Seyfe E, Ching A, Freedman MJ (2018) Riffle: Optimized shuffle service for largescale data analytics. In: Proceeding of the 13th EuroSys Conference, pp 1–15
25. Dong J-Q, He Z-H, Gong Y-Y, Pei-Wen Y, Tian C, Dou W-C, Chen G-H, Xia N, Guan H-R (2022) SMART: speedup job completion time by scheduling reduce task. J Comput Sci Technol 37(4):763–778

Chapter 3
Industrial-Metadata Intelligent Service for Geo-Distributed File System

Abstract Industrial-metadata service of geo-distributed file systems plays an important role in many applications, such as supply-chain management, process mining from log files, and collaboration among business processes deployed over Industrial Internet. The edge computing paradigm provides an ideal infrastructure for some value-added services by using geo-distributed file systems. However, geo-distributed file systems often need an intelligent metadata service mechanism for low-latency data visiting and processing. Unfortunately, it is often a challenge in practice, as the edge computing paradigm often lacks an edge data center that is indispensable for metadata storage and metadata operation. In view of this challenge, an industrial-metadata intelligent service for geo-distributed file system is introduced in this chapter, which includes two layers as user layer and the edge layer, respectively. Industrial-metadata intelligent service is a metadata service designed for this file system. It can sense the user's network latency and dynamically divides the file system directory tree into several subtrees according to the user's access behavior, to reduce the operation latency between the user and the metadata server. By running the intelligent-metadata service, the industrial virtual edge data center can provide high-quality data service for users.

Industrial-metadata service of geo-distributed file systems plays an important role in many applications, such as supply-chain management, process mining from log files, and collaboration among business processes deployed over Industrial Internet. The edge computing paradigm provides an ideal infrastructure for some value-added services by using geo-distributed file systems. However, geo-distributed file systems often need an intelligent metadata service mechanism for low-latency data visiting and processing. Unfortunately, it is often a challenge in practice, as the edge computing paradigm often lacks an edge data center that is indispensable for metadata storage and metadata operation. In view of this challenge, an industrial-metadata intelligent service for geo-distributed file system is introduced in this chapter, which includes two layers as user layer and the edge layer respectively. Industrial-metadata intelligent service is a metadata service designed for this file system. It can sense the user's network latency and dynamically divides the file system directory tree into several subtrees according to the user's access behavior, to

W. Dou et al., *Intelligent Industrial Internet Systems*,
https://doi.org/10.1007/978-981-99-5732-3_3

reduce the operation latency between the user and the metadata server. By running the intelligent-metadata service, the industrial virtual edge data center can provide high-quality data service for users.

3.1 Metadata Service in Geo-Distributed File System

Many industrial applications have distributed deployments of their data resources across the Internet. A distributed file system is frequently used in practice to pilot users to efficiently access the resources deployed in various geographic locations, facilitating across-geography resource management. To highlight a unique large-scale data-resource management among geographically dispersed data centers [1], the distributed file system in this case is frequently referred to as a geo-distributed file system. In order to give users access to large-capacity and highly scalable data storage resources, a geo-distributed file system integrates a number of physical storage resources through the network [2, 3]. Cross-regional data resource access frequently results in high latency because users and servers are frequently dispersed across different regions [4–6].

Metadata is essentially the index data used to describe data resources. It frequently serves as a support for permission control, resource discovery, and storage location indication. Metadata in a geo-distributed file system typically includes details about the file system's structure, data blocks' identification, their physical locations, and access rights. The management of a distributed file system's metadata is crucial to its effective use. Many web-based applications, including supply-chain management, process mining from log files, intelligent aggregation and collaboration among business processes deployed over Industrial Internet, etc., depend heavily on the metadata service of geo-distributed file systems for low-latency data visiting and processing. Despite its small size, metadata plays a crucial role in the distributed file system's overall performance. On a distributed file system, metadata operations make up more than half of all operations. The two methods of managing metadata are frequently distribution management and centralized management. The metadata is kept in a centralized storage node for centralized management. The metadata is divided into a few sub-metadata for distributed management and is then stored across a number of storage nodes. The centralized management is simple to set up and implement, but it is unsuitable for distributed high-load scenarios with low-latency metadata services, especially for mobile equipment's intelligent business processing.

The edge computing paradigm uses a geo-distributed file system in a mobile way to provide an ideal infrastructure for some value-added services as mobile network technologies advance [7]. But for low-latency data visiting and processing, geo-distributed file system-based industrial applications frequently require an intelligent metadata service mechanism. Unfortunately, it is frequently difficult in practice because the edge computing paradigm frequently lacks an edge data center, which is essential for metadata operation and storage. In the paradigm of edge computing, edge servers are deployed globally and primarily serve as computing resources. Compared to a data center, a single-edge server has less storage space. If the distributed edge servers' free storage spaces could be combined to form a virtual edge data center, it might offer both computing power and storage services [8]. By offering both computing and storage services in the same topology space, as opposed to frequently traveling to a remote data center in a different topology space, it could make it easier for applications to run in a low-latency manner [9]. As a result, in the future, more and more mobile applications may be satisfied by intelligent edge computing, which is supported by the metadata service of geo-distributed file systems.

The majority of edge computing environments connect edge servers to geographically dispersed base stations. Through high-speed links, edge servers can communicate with one another and share storage resources [10–12]. Users can connect to an edge server for data transmission within its coverage area. Users can connect to any of the edge servers that cover their location [13]. According to previous research, a user will connect to the closest edge server that covers them to access the data [14–16]. The majority of contemporary file systems use specific metadata management services. Despite being much smaller in size than file data, metadata operations are crucial to overall performance. File systems may be set up on a number of servers in a single data center or in a number of different data centers. Existing file systems frequently perform poorly in terms of metadata access latency when deployed across geographies, failing to provide users with low-latency services because existing metadata services are rarely optimized for latency in cross-geography scenarios.

Figure 3.1 depicts a case study scenario. Three mobile compute nodes are available to handle some model training tasks, and there are five edge servers collectively covering a particular area. To guarantee the integrity of the learning, a lot of data, including the model architecture and training dataset, must be sent to each compute node. An easy illustration is that when a compute node in the learning system moves, it must first obtain the required data from an edge server to begin its training. These data are typically delivered using a centralized cloud server. Although simple to implement, it cannot guarantee effectiveness. Getting these data through the edge servers as they process these crucial data is a promising solution.

Each compute node will connect to the closest edge server in accordance with the settings to request the training data. Figure 3.1a illustrates the way they connect, u_1 will connect to s_1, and u_2 will connect to s_2. The user's location is continually shifting in the mobile edge computing environment. Without sacrificing generality, assume that u_1 removes to $u_1^{'}$, u_2 removes to $u_2^{'}$, and u_3 removes to $u_3^{'}$ after a fixed amount of time. $u_1^{'}$ will connect to s_3, $u_2^{'}$ will stay connected to s_2, and $u_3^{'}$ will connect to s_4 in this circumstance.

It goes without saying that there will be a significant increase in data transfer and retrieval latency when users or compute nodes move. If the transmit hops increase, a transmission delay between two edge servers will inevitably occur. This delay is unacceptable. Furthermore, compared to the scenario shown in Fig. 3.1, real-world scenarios might involve a lot more edge servers and users. The issue will only get

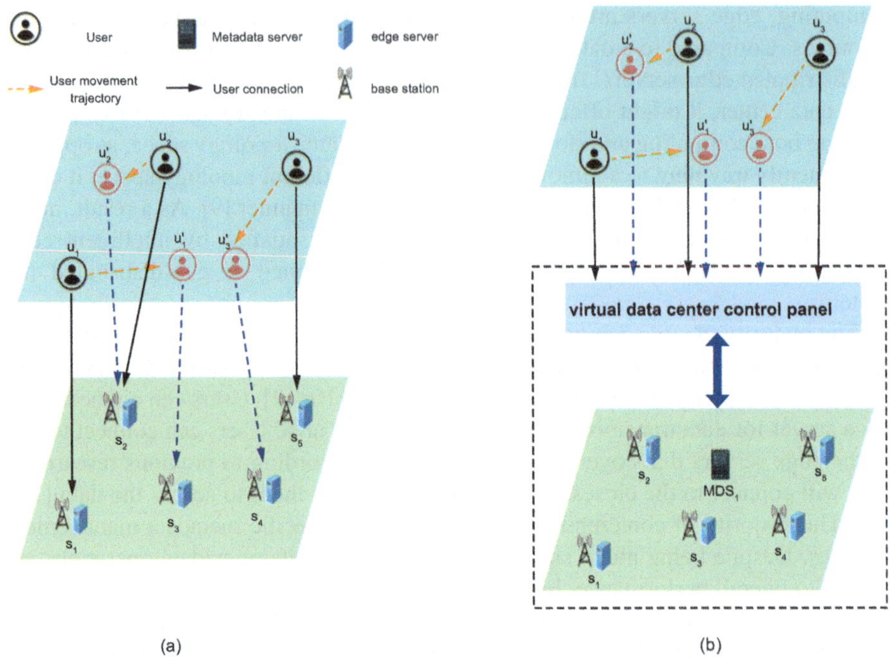

Fig. 3.1 Background of the virtual edge data center in edge computing environment. (a) Traditional edge computing environment. (b) Virtual edge data center in edge computing environment

worse, making it impossible to tolerate the latency brought on by data retrieval and transfer. By combining edge servers, a virtual edge data center is suggested as a solution to this issue. This pattern enables the consolidation of dispersed edge servers into a single entity. The data retrieval time can be effectively controlled by putting a metadata server in each virtual edge data center. Wherever the users are, they only need to communicate with the virtual edge data center control system to request the data they need; the metadata server will then choose which edge server to use to deliver the requested data, as shown in Fig. 3.1b. The overall system latency can be effectively managed in this way.

3.2 Industrial Distributed Virtual Edge Data Center

All nodes in an industrial virtual edge data center can access a single and shared namespace by consolidating storage space on various edge servers to create a distributed file system. There are some similarities and some differences between the virtual concept in this chapter and the virtual concept in cloud computing. First, many virtual data nodes are deployed in edge servers and kept apart, similar to virtualization technology. Second, users can view metadata service operations transparently, such as data retrieval from virtual data nodes and subtree migration to mobile data station (MDS). Users can upload or download the data they require by connecting to edge servers in this architecture without having to own all of the system's nodes. Geographically dispersed throughout various regions, edge servers appear to the user to be a full-fledged data center. A controller is placed close to an MDS within a constrained geographic area. There will be several areas split up in a very large geographic area, and controllers will be placed in each area.

Definition 1 (Industrial virtual edge data center) *A data-intensive job that runs in the edge computing environment can be divided into numerous smaller tasks,* $T = \{t_1, t_2, \ldots, t_n\}$. *A group of edge servers will be used to deploy each of these subtasks* $S = \{s_1, s_2, \ldots, s_m\}$. *The capacity of an edge server's idle storage is* sc_i. *If certain subtasks regularly accessed a faraway cloud data center, it commonly resulted in excessive latency for the execution of those subtasks. This observation suggests that aggregating these unused storage capacities into a virtual resource pool dynamically may be an alternative method. We'll approach this virtual resource pool like a fictitious edge data center. The virtual edge data center's storage capacity is:* $\sum_{i=1}^{i=m} \mathrm{sc}_i, \forall \mathrm{sc}_i \in \mathrm{SC}$.

A collection of edge servers can be seen as a virtual edge data center in the data deployment architecture, which can then be seen as a single data center. The virtual edge data center has three fundamental characteristics:

- Dynamical: A dynamic data center with its own life cycle is the virtual edge data center. The capacity and life cycle of a virtual edge data center are constantly changing because they are produced by an unfixed set of edge servers.
- Task-driven: A specific data-intensive task drives a virtual edge data center. The virtual edge data center's life is over if the task has been completed.
- Service-oriented: A virtual edge data center has all the basic capabilities of a data center, including its own retrieval system and the ability to offer users a metadata service.

A grouping of r_j covered edge servers can be thought of as an industrial virtual edge data center, which can be thought of as a single data center. Each virtual edge data center will have a metadata server to offer data retrieval services. The geo-distributed file system that powers the intelligent metadata service operates. This file system's servers play the three functions of *Manager*, *MDS*, and *Virtual Datanode*. *Users* receive metadata services from *MDS* and *Manager*. The geo-distributed file system model for the industrial virtual edge data center is shown in Fig. 3.2.

A set of nodes, usually three, that are installed on different edge servers and are used to execute the Manager and ZooKeeper applications. The manager is used to monitor each MDS's condition, particularly when MDS failures occur. It is intended to serve as the file system's chief node. MDSs are responsible for maintaining the metadata for the file system. The MDS nodes are located either in the data center or on the edge server in terms of location. A user can access any metadata server to start

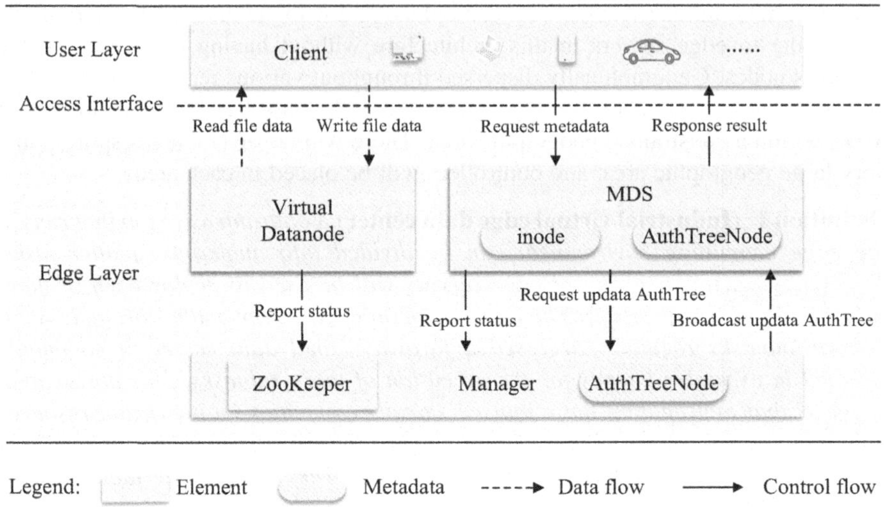

Fig. 3.2 The virtual edge data center file system framework

using the Intelligent Metadata Service. Centers for file data processing and storage are virtual data nodes. In edge computing, virtual data nodes are dispersed throughout a number of edge servers, with each edge server having the potential for one or more virtual data nodes. A distributed system consists of several nodes connected through a network. Both nodes and networks are susceptible to failure at some time. In the case of a node or network fault, the system must be able to continue providing services to users. Availability describes a system's ability to function continuously. The high availability of Manager and MDS components in the scheme described in this chapter significantly contributes to the high availability of metadata services. In order to accommodate application situations with high availability requirements, the framework uses the following strategies to increase service availability.

For one thing, to maintain high availability for both the Manager and the MDS (Metadata Server) in the system, the following strategies are implemented. Three nodes are used to execute the Manager application, but only one of them is now in use; the others are standby nodes. Write operations to the Manager AuthTree must be finished on at least two Manager nodes before being forwarded to Zookeeper. When the active Manager node fails, Zookeeper alerts the backup node that the Active Manager node is unavailable. Any standby node that becomes active registers with Zookeeper. This method ensures that the Manager's functions are still available in the case that up to one Manager node fails. To maintain high availability for the MDS (Metadata Server) component, the following measures are taken. There are three copies of an inode's metadata for a subtree, and each one is preserved by a distinct MDS. One MDS is the major copy, while the other two MDSs are the secondary copies. The write action must be completed on at least two replicas before the Client is able to receive a valid commit response. If the MDS serving as the primary copy fails, the Manager upgrades the first secondary copy to the primary

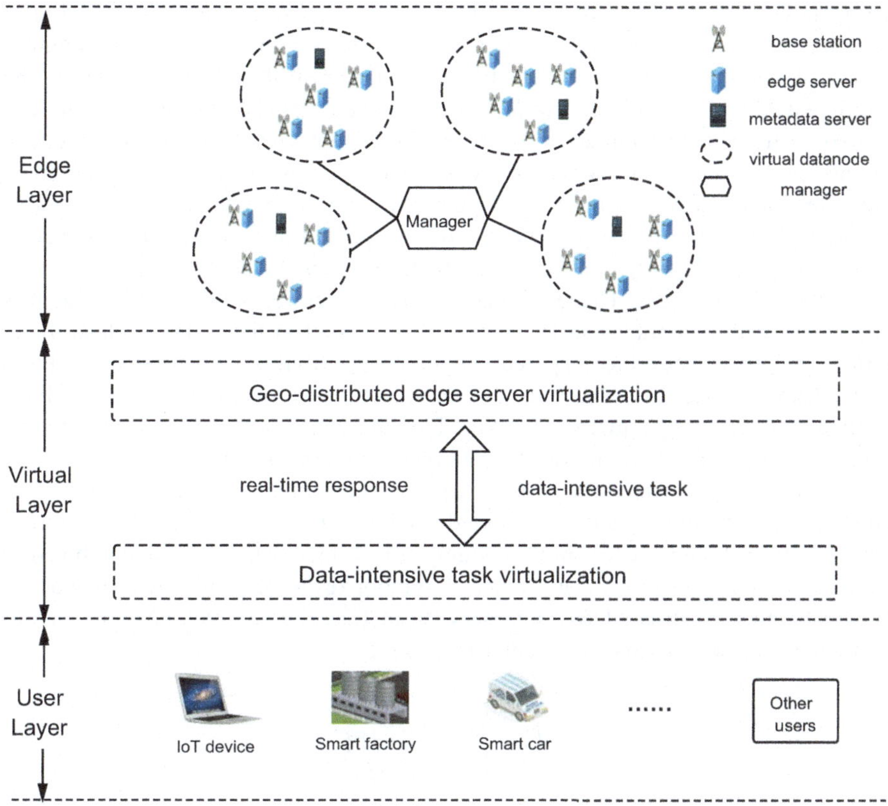

Fig. 3.3 Architecture of a virtual data center in edge computing environment

copy. If up to one MDS copy fails, this approach can ensure the availability of inode metadata for a subtree.

Figure 3.3 shows the architecture of the virtual edge data center in the edge computing environment. In a modern metropolis, a huge number of computer tasks are continuously created. IoT devices, autonomous vehicles, industrial manufacturing systems, traffic management systems in cities, and many more things demand low-latency computing resources. By positioning edge servers close to the base stations, the edge computing paradigm is able to handle the task's computational demand. In general, even though computing-driven operations typically require storage, edge servers are able to offer both processing capabilities and substantial storage capacity. This virtual edge data center can respond to requests in real time and can function reliably in an edge computing setting. It is possible to think of the various virtual edge data centers that have been set up or are currently in use as a file system.

The two layers that make up this architecture's essential parts are the user layer and edge layer, as depicted in Fig. 3.3. A geo-distributed file system is set up on the virtual edge data center to make it simpler for users to query for the required data. Running the Intelligent Metadata Service on the file system will increase communication between the user layer and edge layer and bring them closer together. The Intelligent Metadata Service, a wide-area networked metadata service, runs in data centers scattered across numerous locations. Intelligent Metadata Service makes use of the users' local data access to cut down on access latency. The geographic environment of many applications affects how they access data. In other words, users close by favor accessing comparable material. For instance, the navigation system of a vehicle only consults high-precision maps of the neighborhood rather than other distant areas. Using a directory tree structure, data in a file system is organized, with essential data frequently being stored in the same or close directories. The file system directory tree is dynamically divided into subtrees by Intelligent Metadata Service based on an analysis of user operation behavior. Each subtree is managed by a metadata server. The subtree is positioned to get the fewest access latency requests in the future from incoming access requests.

Given a group of edge servers $S = s_1, s_2, \ldots, s_n$ from a particular region A. Each edge server's storage capacity corresponds to the set $C = c_1, c_2, \ldots, c_n$. The storage capacity of each edge server in the vicinity A of a metadata server is combined to form the virtual micro edge data center. A symbolizes a location's area, such as the central business district of Beijing. Because of this, the storage capability of this particular industrial virtual data center is:

$$\mathrm{SC} = \sum_{i=1}^{i=n} c_i, \forall c_i \in C$$

For each u_j will have a sequence of request data. Use $r_1, r_2, \ldots, r_m$ to represent data requests of each user. Given a set of user $U = u_1, u_2, \ldots, u_m$. The total requests in the region A are:

$$R = \sum_{j=1}^{j=m} r_j, \forall u_j \in U$$

To create a distributed file system that all nodes in the virtual edge data center can access, a virtual edge data center can combine storage space from various edge servers. The virtual notion and the virtual concept in cloud computing have some parallels and some differences. First, using virtualization technology, numerous virtual data nodes are put in edge servers and kept isolated. Second, users can see openly processes performed by the metadata service, such as subtree migration to MDS and data retrieval from virtual data nodes. In this architecture, users can upload or download the data they need by connecting to an edge server without owning all of the system's nodes. The user perceives edge servers, which are geographically scattered across several areas, as a full-fledged data center. Within a limited geographic area, a controller is positioned close to an MDS. A very vast geographic area will be divided into multiple areas, with controllers located in each area. The user layer often controls the virtual edge data center in this design.

Manager: The other Manager nodes work as a backup in the event of a single point of failure, and only one Manager node is active at any given time. If the active Manager node fails, ZooKeeper re-selects a standby node as the new active node. The active Manager node monitors the performance of each MDS. The moment the MDS fails or goes offline, the active Manager node will notice it and handle it. The proposed file system's stability and security depend on ZooKeeper, which enables data consistency between manager nodes.

Metadata Server: In Intelligent Metadata Service, there are two types of metadata. *INode*, which describes the directory structure and file information, is the initial type of metadata. Each MDS maintains and persists only the inodes present in its managed subtree. Intelligent Metadata Service uses RocksDB, an embeddable persistent key-value store for fast storage, to permanently store metadata on disk. Every time a user creates or modifies a file, they must write the inode on the MDS that manages the file's metadata. When writing an inode, RocksDB changes the inode data in memory after appending the operation to the log file on the disk. The second kind of metadata is called *AuthTreeNode*, which explains how the file system directory tree is divided up and which MDS is in charge of each subtree. Users can find out which MDS is in charge of a directory or file by querying the *AuthTreeNode* for a path. Only during migration operations is the *AuthTreeNode* updated. All MDSs and managers have replicas of the *AuthTreeNode*. The primary replica of the *AuthTreeNode* is the replica on the manager, and any updates to the primary replica are broadcast to all replicas via ZooKeeper, maintaining the consistency of the *AuthTreeNode* data throughout the system. The dynamic subtree partition, which is managed by a separate MDS, divides the file system directory tree into a number of subtrees. In the Intelligent Metadata Service, MDSs keep track of user actions and gather information on latency between users and MDSs.

Virtual data node: The idea of virtual is similar to that of the dual-meaning virtual edge data center. These virtual data nodes are initially divided in the edge server, much like virtualization technology, and do not normally interface with one another. As a result, virtual data nodes are transparent to users due to the transparent processes performed to retrieve data from them. In other words, users query MDSs to locate the data they require on an edge server, download it in their thoughts, and save it on the edge server, but the data is really stored and processed by virtual data nodes.

User: To get a list of current online MDSs, a user can connect to any known MDS. The user checks the latency between each MDS and itself on a regular basis and reports this data to all MDSs that have already established connections with the user. In an edge computing environment, the user will naturally want to communicate with the closest MDS based on geographic proximity to benefit from reduced latency and acquire information as soon as possible. When a user uses a path, it first checks to see if it has used the path before. If it has, the user is taken straight to the MDS connected to that path. If not, the user establishes a connection with the MDS closest to them and inquires as to which MDS is in charge of the directory or file that path denotes.

3.3 Intelligent Metadata Service for Industrial Edge Data Center

The intelligent metadata service (IMS) is a distributed service that operates across multiple MDSs.

The underlying assumption of the intelligent metadata service is that user access behavior to metadata adheres to the locality principle, including temporal and spatial locality. Making sure people get the info they need in a timely manner is crucial. Intelligent metadata service chooses a directory subtree controlled by this MDS and migrates it to the suitable MDS, often the closest MDSs to the user, when the MDS determines that the user operation latency with it exceeds the threshold T_t. As a result, not only will the target MDS not be overloaded due to the removal of far and nearby users, but the user's operation latency will also be kept as low as feasible.

Latency refers to the amount of time that elapses between when a user sends a request and when the user receives a response. First, define the latency threshold as follows:

Definition 2 (latency threshold) *Set a time limit T_t above which the directory subtree in this MDS may change if the latency between a user and its destination MDS exceeds T_t. The time value T_t is referred to here as the latency threshold.*

The reduction of processes whose latency exceeds the threshold is the aim of the intelligent metadata service, to define intelligent metadata service's objective.

$$p_{\text{latency}} = \frac{\text{Number of operations with latency} < T_t}{\text{Number of operations}}$$

where p_{latency} is the percentage of operations among all operations whose latency is below the latency threshold. The goal of intelligent metadata service can be expressed as

$$\text{Maximize}\left(p_{\text{latency}}\right)$$

The best way to accomplish this goal is to migrate the client-accessible metadata inode$_x$ to the MDS that is nearest to the client. However, it is impossible to foresee every operation in a real file system with complete accuracy.

A user is more likely to access a path or one of its nearby paths in the near future if they have just recently accessed it, according to the storage system. There is a strong possibility that a user will frequent a path again in the near future if they have visited it frequently recently. Since a user frequently engages in activities within a specific area where the mobile data station (MDS) is located, there is a higher probability of the MDS being selected as the target MDS. This phenomenon holds particular significance in the context of edge computing.

Intelligent metadata service benefits from the user's ability to access the file system locally. In order to detect the migration of and its subdirectories when timeout access to inode_i happens, intelligent metadata service examines the operation behavior of users who visit inode_i in a time frame.

Step 1. Identify inodes that need to be optimized. Put the inodes that have been accessed during this operation into a queue if the latency of the operation exceeds T_t. For queue order optimization, inodes are removed from the queue.

mds_i records the action of inode_x from user_c and checks L_c^i immediately. Network latency between user_c to mds_i is L_c^i. Push the pair $\langle \text{inode}_x, \text{time}_{\text{now}} \rangle$ into Q_i if $L_c^i > T_t$. Q_i indicates the list of inodes in the queue that are awaiting optimization in mds_i. time_{now} indicates the current time. INodes in Q_i are not optimized immediately when they are pushed into Q_i. An unintended timeout log could exist. INodes in Q_i must wait till there is adequate time to collect more logs in order to perform better optimization. Try to identify the ideal target MDS for each inode_x in Q_i for which you have waited long enough.

Step 2. Find the best target MDS for the INode. Locate the ideal MDS to use as the migration target for the removed INode. If this doesn't work, try using breadth-first search to discover the best target MDS for the currently selected INode's child nodes.

Algorithm 3.1 describes the process to find the best target MDS for inode. Here, mds_c denotes current mds which running this algorithm. $ep_{\text{inode}_t}(\text{mds}_i)$ means that if inode_t is in mds_i, the estimated proportion of operations access $inode_t$ whose latency $<T_t$. $p_{\text{latency}}^{\text{expect}}$ denotes a threshold p_{latency} expected to be achieved.

The procedure to identify the ideal target MDS for inode is described in Algorithm 3.1. mds_c denotes current MDs running this method. $ep_{\text{inode}_t}(\text{mds}_i)$ signifies thatthe estimated percentage of operations whose delay $<T_t$ if inode_t is in mds_i. $p_{\text{latency}}^{\text{expect}}$ indicates a target value p_{latency} that should be reached.

Algorithm 3.1: The Process to Find the Best Target MDS

Require: *inode*: The INode needs to optimize; MDS_list: The list of all MDS;
Ensure: state: a boolean value represents whether found target mds or not; $inode_t$:the root of subtree to be partitioned; mds_t: the migrate target mds;

```
q.push(inode) // q denotes the queue used for breadth-first search.
while True do
    inode_t = q.pop();
    if time_now − time_lo^{inode_t} then
        l = list of users accessed inode_t during [time_now − T_tr, time_now)
        count_{user_x} = the access time from user_x during [time_now − T_tr, time_now)
        time_lo^{inode_t} = time_now
        if l is empty then continue;
        for mdsi in MDS_list do
            in_threshold_req = 0;
            for userj ∈ l do
                if L_j^i < Tt then in_threshold_req+=count_{user_j}
            ep_{inode_t}(mds_i) = in_{threshold_req} / Σ_{user_t ∈ l} count_{user_t}
        mds_t = the mds which has the largest ep_{inode_t}
        if epinodet (mdst ) > p_latency^expect then return (True, inode_t, mds_t)
```

(continued)

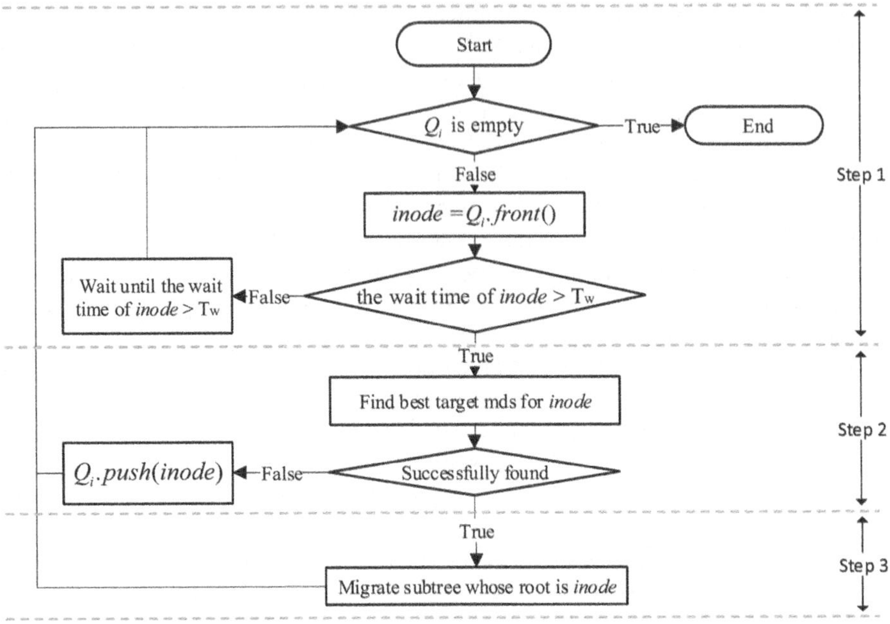

Fig. 3.4 The flow chart of subtree partition and migration method in intelligent metadata service

```
16:      else q.push(childnodes of inodet )
17: return False
```

In Algorithm 3.1, T_{tr} refers to the operation logs' traceback time. Users typically access files in a file system from top to bottom in accordance $inode_x$with the directory hierarchy. Users typically need to access one inode in a subtree before accessing another inode. Consequently, a rough estimate of the subtree can be made. The estimate of the subtree is equivalently improved if a suitable MDS can be found to use as a target for migration, which enhanced of $p_{latency}$. A breadth-first search will be done to discover a suitable migration target for other nodes in the subtree if a suitable MDS cannot be found for the migration target $inode_x$. The root node to split a new subtree will be the $inode_z$ if a suitable migration target is located. The destination MDS will receive this subtree's migration.

Step 3. Migrate subtree to target MDS. The INode whose root has been determined to be the best target MDS is the root of the partition, which is a subtree of the current directory trees. Move this subtree to the desired MDS after that.

Figure 3.4 shows the flow chart of subtree partition and migration method in intelligent metadata service which is separated into 3 steps.

In Fig. 3.5, the migration procedure is shown. If a failure prevents an MDS from receiving broadcast information, it will receive the most recent "UpdateAuthTree" message once everything are back to normal. All missing update transactions will be retrieved by the MDS, which will then perform these operations again. This

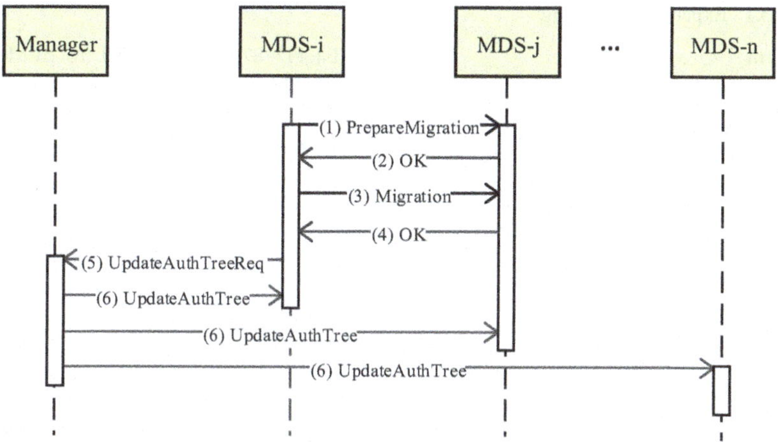

Fig. 3.5 Sequence diagram of migration

guarantees that the system is consistent. As a result, only one MDS will ever be in charge of managing an INode.

3.4 Analysis of Industrial-Metadata Intelligent Service

An access log dataset is used to test the read-write latency and throughput of intelligent metadata service and Hadoop Distributed File System (HDFS), and the first 100,000 access records from the dataset are evaluated to evaluate the performance of industrial-metadata intelligent service. Each IP address in the dataset is associated with a particular region.

End-user devices (smartphones, tablets, and laptops) and IoT devices can typically reach an edge server with a latency of less than 10 ms. As a result, if an operation's latency is less than 10 ms, it can be considered completed at the edge server in the same area as the user. In the experiment, set T_t in intelligent metadata service to 10 ms. In a virtual edge data center, different regions are usually close to each other.

Assume that there are five, seven, nine, eleven, thirteen, and fifteen regions in the industrial virtual edge data center. When using intelligent metadata service, each area has its own metadata server. The entire virtual edge data center has a single metadata service when employing single region metadata services, like the HDFS metadata service. The experiments set the latency range between various nodes in the virtual edge data center to 1–60 ms, 1–80 ms, and 1–100 ms, which corresponds to the delay between various data centers in reality. The network delay between regions is selected at random from the delay interval for each group of comparative tests, along with a number of regions and a delay interval. The setup is depicted in Table 3.1.

Table 3.1 Experimental setting

Platform	Intel i7–8700 6 Core CPU, 16GB Memory
Dataset size	100,000 access logs
Number of regions	5, 7, 9, 11, 13, 15
Latency range	1–60 ms, 1-80 ms, 1–100 ms
T_t	10 ms

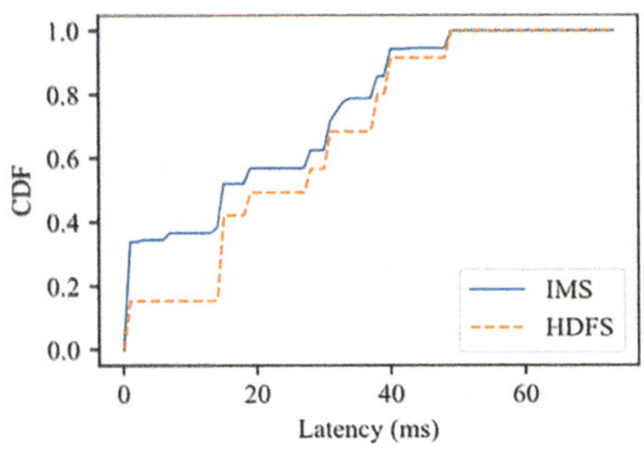

(a)Write, max latency < 60ms

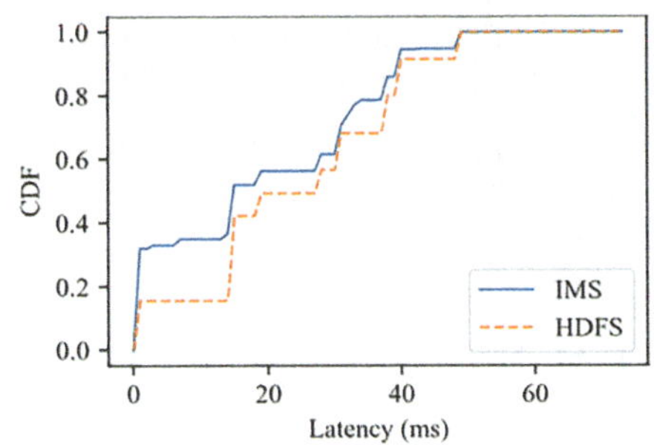

(b) Read, max latency < 60ms

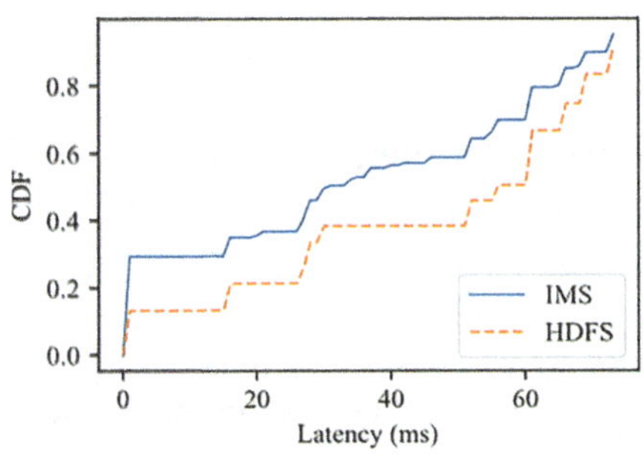

(c)Write, max latency < 80ms

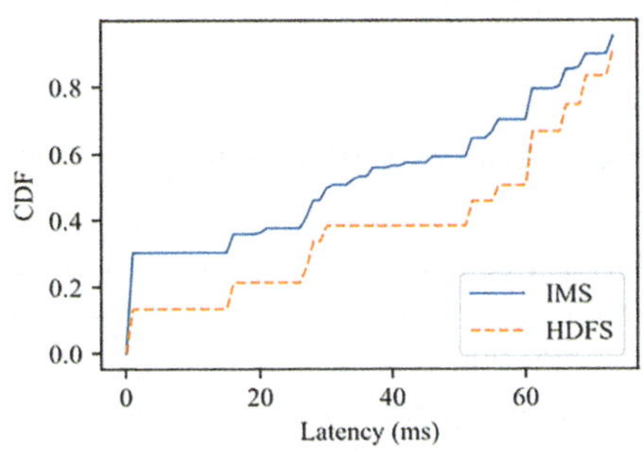

(d) Read, max latency < 80ms

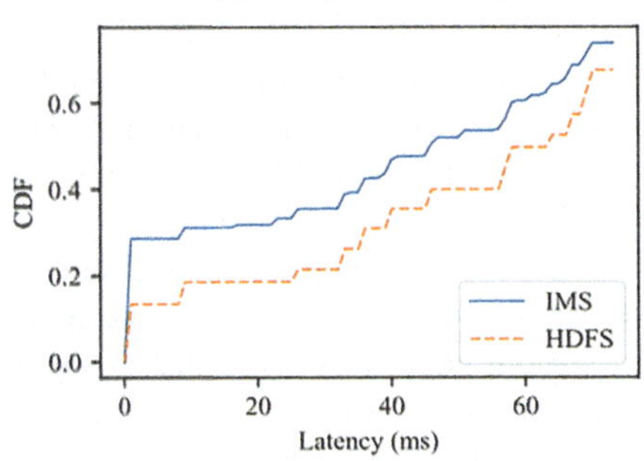

(e)Write, max latency < 100ms

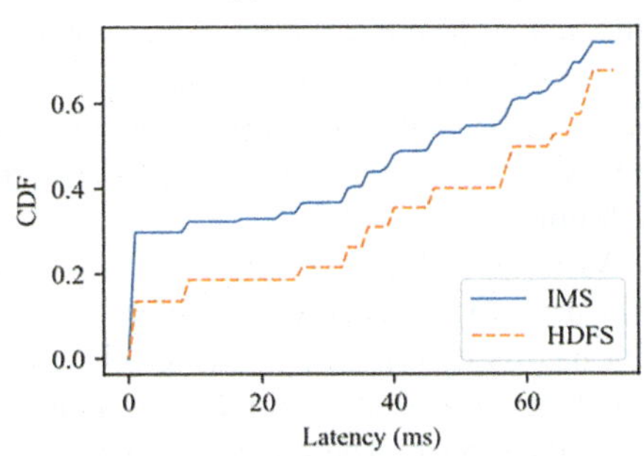

(f) Read, max latency < 100ms

Fig. 3.6 Cumulative distribution function (CDF) of operation latency. (**a**)Write, max latency <60 ms. (**b**) Read, max latency <60 ms. (**c**) Write, max latency <80 ms. (**d**) Read, max latency <80 ms. (**e**)Write, max latency <100 ms. (**f**) Read, max latency <100 ms

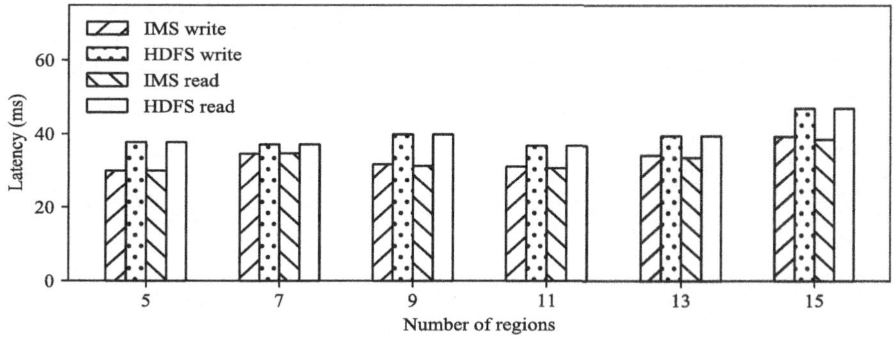

Fig. 3.7 Average latency of IMS and HDFS

Latency evaluation. Figure 3.6 shows the cumulative distribution function of operation latency in intelligent metadata service and HDFS.

Figure 3.6 depicts the cumulative distribution of operation latency in IMS and HDFS for an industrial virtual edge data center operating across five regions. IMS is 135% quicker than HDFS when inter-node latency is less than 60 ms, with 36.8% of write operations finishing within 10 ms compared to 15.65% of write operations finishing within 10 ms for HDFS. When the inter-node latency is less than 80 ms, 29.5% of IMS write operations are finished in 10 ms, compared to just 13.55% of HDFS write operations, making IMS 118% faster. With a latency of less than 100 ms between nodes, 32.3% of IMS write operations are finished within 10 ms, compared to 18.7% of HDFS write operations, a 72% improvement for IMS over HDFS. The distribution of latency for read and write operations is similar.

When compared to HDFS, IMS can considerably reduce latency, enabling more operations to be executed with reduced latency. In terms of operation latency, IMS outperforms HDFS for inter-node latency ranges of 1–60 ms, 1–80 ms, and 1–100 ms. The majority of tasks may be finished in a matter of seconds, which enables programs to run more quickly and enhance user experience.

IMS and HDFS average latency in various regions are shown in Fig. 3.7. According to the results of the experiment, the number of regions has no impact on the typical operation latency of IMS. When the number of regions is set to 5, 7, 9, 11, 13, and 15, IMS's average operation latency is much lower than HDFSs.

Both HDFS Namenode and intelligent metadata service MDS append metadata write operations from tasks first to the disk's log file. The information then updated in memory. They then send a confirmation to the client. Processing time on the server and network transmission time make up the two halves of an operation's delay.

Figures 3.8 and 3.9 depict the operation per second (OPS) of write and read operations in intelligent metadata service and HDFS, respectively. Intelligent metadata service significantly reduces latency, which also helps to improve system throughput.

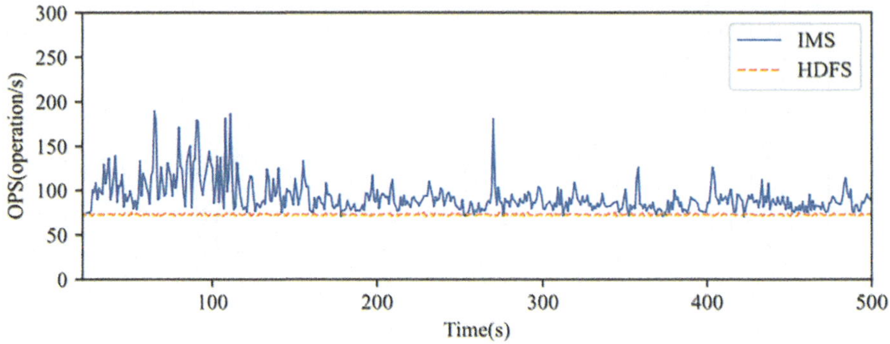

Fig. 3.8 Operation per second (OPS) of write operation (5 regions)

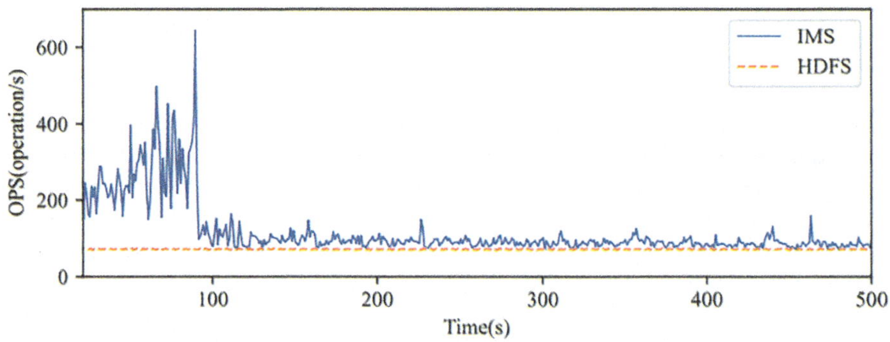

Fig. 3.9 Operation per second (OPS) of read operation (5 regions)

By drastically reducing latency, intelligent metadata services boost system throughput. For IMS, the average write throughput is 75.01 ops, compared to 55.09 ops for HDFS. IMS and HDFS both have average read throughputs of 55.09 ops and 79.28 ops, respectively. IMS has a write throughput of 136% and a read throughput of 144% of HDFS. In the trials, HDFS throughput is more consistently smooth than IMS, which is more prone to fluctuations. This is due to the fact that IMS dynamically modifies subtree partitioning based on user access while it is running, which causes throughput to change. The experimental findings show that IMS has a higher throughput than HDFS.

In a file system, data is arranged in a directory tree structure, and similar data is typically put in the same or nearby directories. The intelligent metadata service dynamically creates subtrees within the file system directory tree based on an analysis of user operation behavior. Each subtree is controlled by a metadata server. In order to process as many incoming access requests as feasible quickly and reduce access latency, the subtrees are situated on the proper nodes. The centralized metadata management method used by HDFS, in contrast, routes all metadata

Table 3.2 Comparison between IMS and metadata service of HDFS (5 regions)

Item	IMS	Metadata service of HDFS
Average latency	45.51 ms	56.69 ms
Average throughput	77 operation/s	55 operation/s
Task-driven	Yes	No
Cross region support	Yes	No

queries to a single node for processing. As a result, IMS performs better than HDFS in terms of latency and throughput.

The evaluation's findings support the aforementioned assertion. The experimental findings of the examination of latency show that intelligent metadata service performs much better than HDFS. This demonstrates how intelligent metadata service can perform more operations with less delay. This highlights how the intelligent metadata service can handle more operations in a shorter amount of time. The outcomes of single region metadata services and IMS are contrasted in Table 3.2. For file systems in geo-distributed environments, intelligent metadata service has advantages in terms of latency and throughput.

3.5 Summary

In this chapter, an industrial-metadata intelligent service for geo-distributed file system is introduced, which including two layers as user layer and the edge layer, respectively. To make full use of the storage capacity, a geo-distributed file system is deployed on the edge layer, covering several geographic regions. There is a metadata server deployed in each region to manage the metadata of the file system so that users can query it for requested data. Industrial-metadata intelligent service is a metadata service designed for this file system. It can sense the user's network latency and dynamically divides the file system directory tree into several subtrees according to the user's access behavior, so as to reduce the operation latency between the user and the metadata server. By running an intelligent metadata service, the industrial virtual edge data center can provide high-quality data service for users. The experimental results show that this novel architecture works well and the proposed metadata service can reduce the access latency.

References

1. Tao V, Shapiro M, Rancurel V (2015) Merging semantics for conflict updates in geo-distributed file systems. In: Proceedings of the 8th ACM International Systems and Storage Conference, pp 1–12
2. Huang Y, Shi Y, Zhong Z, Feng Y, Cheng J, Li J, Fan H, Li C, Guan T, Zhou J (2019) Yugong: geo-distributed data and job placement at scale. Proc VLDB Endowment 12(12):2155–2169

3. Wang H, Shen H, Li Z, Tian S (2021) GeoCol: a geo-distributed cloud storage system with low cost and latency using reinforcement learning. In: 2021 IEEE 41st International Conference on Distributed Computing Systems (ICDCS), pp 149–159
4. Liu K, Peng J, Wang J, Liu W, Huang Z, Pan J (2020) Scalable and adaptive data replica placement for geo-distributed cloud storages. IEEE Trans Parallel Distributed Syst 31(7): 1575–1587
5. Bogdanov KL, Reda W, Maguire Jr GQ, Kostić D, Canini M (2018) Fast and accurate load balancing for geo-distributed storage systems. In: Proceedings of the ACM Symposium on Cloud Computing, pp 386–400
6. Oh K, Qin N, Chandra A, Weissman J (2019) Wiera: policy-driven multi-tiered geo-distributed cloud storage system. IEEE Trans Parallel Distributed Syst 31(2):294–305
7. Liu B, Jiang X, He X, Qi L, Xiaolong X, Wang X, Dou W (2022) A deep learning-based edge caching optimization method for cost-driven planning process over IIoT. J Parallel Distributed Comput 168:80–89
8. Xia X, Chen F, Grundy J, Abdelrazek M, Jin H, He Q (2021) Constrained app data caching over edge server graphs in edge computing environment. IEEE Trans Serv Comput 15(5):2635–2647
9. Xia X, Chen F, He Q, Grundy JC, Abdelrazek M, Jin H (2020) Cost-effective app data distribution in edge computing. IEEE Trans Parallel Distributed Syst 32(1):31–44
10. Liu B, Xiaolong X, Qi L, Ni Q, Dou W (2021) Task scheduling with precedence and placement constraints for resource utilization improvement in multi-user MEC environment. J Syst Archit 114:101970
11. Meng J, Tan H, Xu C, Cao W, Liu L, Li B (2019) Dedas: online task dispatching and scheduling with bandwidth constraint in edge computing. In: IEEE INFOCOM 2019-IEEE Conference on Computer Communications, pp 2287–2295
12. Lin L, Liao X, Jin H, Li P (2019) Computation offloading toward edge computing. Proc IEEE 107(8):1584–1607
13. Ren Y, Leng Y, Cheng Y, Wang J (2019) Secure data storage based on Blockchain and coding in edge computing. Math Biosci Eng 16(4):1874–1892
14. Liu B, Meng S, Jiang X, Xiaolong X, Qi L, Dou W (2021) A QoS-guaranteed online user data deployment method in edge cloud computing environment. J Syst Archit 118:102185
15. Xu X, Huang Q, Zhu H, Sharma S, Zhang X, Qi L, Bhuiyan MZA (2020) Secure service offloading for internet of vehicles in SDN-enabled mobile edge computing. IEEE Trans Intell Transp Syst 22(6):3720–3729
16. Wang L, Jiao L, He T, Li J, Bal H (2020) Service placement for collaborative edge applications. IEEE/ACM Trans Networking 29(1):34–47

Chapter 4
Intelligent Cooperation Technology for Multi-Access Device

Abstract Due to the complex environment changes and obstacles caused by the rapid movement, the network state of the user device is often unstable. A large number of interruptions and resolution reductions seriously affect the user experience. Through direct transmission between devices, device-to-device (D2D) can effectively improve the cooperation quality of service. Accurate decision for helper device selection is the key to optimize services based on D2D technology. However, the existing methods through D2D rarely consider the states and the intention of the helper devices when selecting cooperative devices. This is highly probable to reduce the efficiency of D2D transmission and the cooperation enthusiasm of helper devices. In view of these challenges, an intelligent cooperation technology for multi-access device is introduced in this chapter. Algorithms and criteria for selecting the best devices for cooperation are described, and dynamic decision-making algorithms for multi-path parallel transmission are also discussed. A dynamic smart decision method for device selection in D2D-assisted multi-path video transmission (DS-DAMP) network is proposed. Consider the status and intention to select the appropriate helper devices to improve the satisfaction of helper devices (HDs) and the video quality of requesting devices (RDs).

Due to the complex environment changes and obstacles caused by the rapid movement, the network state of the user device is often unstable. A large number of interruptions and resolution reductions seriously affect the user experience. Through direct transmission between devices, device-to-device (D2D) can effectively improve the cooperation quality of service. Accurate decision for helper device selection is the key to optimize services based on D2D technology. However, the existing methods through D2D rarely consider the states and the intention of the helper devices when selecting cooperative devices. This is highly probable to reduce the efficiency of D2D transmission and the cooperation enthusiasm of helper devices. In view of these challenges, an intelligent cooperation technology for multi-access device is introduced in this chapter. Algorithms and criteria for selecting the best devices for cooperation are described, and dynamic decision-making algorithms for multi-path parallel transmission are also discussed. A dynamic smart decision method for device selection in D2D-assisted multi-path

W. Dou et al., *Intelligent Industrial Internet Systems*,
https://doi.org/10.1007/978-981-99-5732-3_4

video transmission (DS-DAMP) network is proposed. Consider the status and intention to select the appropriate helper devices to improve the satisfaction of helper devices (HDs) and the video quality of requesting devices (RDs).

4.1 D2D Cooperation in Multi-Access Transmission Network

Video streaming applications have gradually risen to the top of the most popular applications list thanks to the quick development of mobile cloud computing and multimedia computing as well as the widespread use of mobile devices. The Industrial Internet of Things (IIoT), live and on-demand video in our daily lives, online video courses in education, self-driving cars in smart cities, and other video-based services all have significant growth potential [1, 2]. By 2022, there will be 5.5 billion mobile users worldwide, and 79% of that will be mobile video traffic. In the meantime, the IIoT is characterized by frequent instability in the network status of the user device on the transport due to user movement, complicated environment changes, and obstacle occlusion [3]. Numerous interruptions and resolution drops have a negative impact on the user's experience watching videos. Device-to-device (D2D) communication is seen as a potential technique for resolving the aforementioned problems through direct device communication [4, 5].

Device-to-device (D2D) communication allows two devices to talk to each other without going through the core network [6]. The advantages of this essential candidate technology for the traditional cellular network include offloading wireless traffic, increasing system throughput, lowering communication delay, enhancing spectrum and energy efficiencies, and more [7]. D2D communication is additionally thought to be a way to improve the performance of edge customers, who ordinarily receive poor services, and increase the coverage area of the cellular network [8]. D2D communication is affected by a number of factors, such as processing efficiency, device transmit and receive power, spatial signal interference, data transfer distance, etc. D2D communication pairs can use little power and high transmission rates to send data over stable distances for extended periods of time [9]. However, the performance and effectiveness of D2D communication will be adversely impacted by the ongoing distance fluctuation and the change from a single-hop to a multi-hop neighbor connection [10]. In general, the single-hop neighbor relationship's stability is essential for D2D communication [11]. As a result, the level of cooperative device selection decision-making is essential for high-quality video services using D2D communication.

A lot of research focus on D2D-based video streaming optimization. The existing approaches rarely consider the current states and cooperative intentions of the helper devices, such as the watching status, the state of charge, the state of the cache, and the connection quality, when choosing cooperative devices. Instead, they only consider if the connection of the assistive devices is accessible. Unplanned caching, poor caching performance, and persistent low battery warnings may have a negative influence on cooperating device satisfaction.

The majority of modern mobile devices also provide various network interfaces, such as WiFi, Bluetooth, and 4G or 5G cellular network interfaces [12–15]. Modern mobile devices, in particular, frequently enable dual card and dual standby, allowing two communication cards in a single device to concurrently consume network services from multiple network providers. Since D2D communication is often managed in a cellular manner, the devices can communicate directly by recycling cell resources or by using the dedicated D2D spectrum. When the terminal device wishes to utilize different networks, in parallel transmission, it is not possible to use multiple network cards for multiple connections; rather, they can only be modified by actively switching between various networks. For instance, the mobile user is using WiFi on their phone to view the video. When the WiFi network is down, videos won't instantly convert to loading over a 4G cellular network. Instead, they are restricted to using a 4G cellular network to reload video resources when the user deliberately disconnects from WiFi. This approach of connection setup, which relies on a single network interface, is unable to make use of numerous network interfaces or fully take advantage of D2D collaboration.

4.2 Device Selection During Cooperative D2D Transmission

The network state of the device is different on a high-speed transport because of the impact of the complex environment and the variations in hardware. D2D technology has been demonstrated to be a practical optimization method. By creating a direct link between devices for data transmission, video quality is ensured on both ends of the device, enabling high-speed data exchange. Many individuals enjoy watching films while riding the bus, tram, or train. These users will have relatively device connections and relatively fixed locations. In addition, whenever there are video services available, such as live broadcasts of important athletic events, live broadcasts of games, and live broadcasts of concerts, numerous video users can easily generate the same video service requests. The aforementioned situation is therefore suitable for starting a D2D connection between users to enable resource sharing, reciprocal transmission, and high-quality video viewing.

However, current D2D-based video stream optimization techniques usually only consider whether the connection between devices is available, and rarely consider the current state of the cooperating devices and the expected level of cooperation, such as the watching status, the state of charge, the state of the cache, and the connection quality. The unpredictable behavior and poor connection quality will reduce the efficiency of D2D transmission. The satisfaction of collaborating devices may be significantly impacted by unfulfilled user cache intentions, poor cache performance, and frequent power shortage notifications, which will reduce the efficiency of D2D transmission and the motivation of helper devices for cooperative behavior.

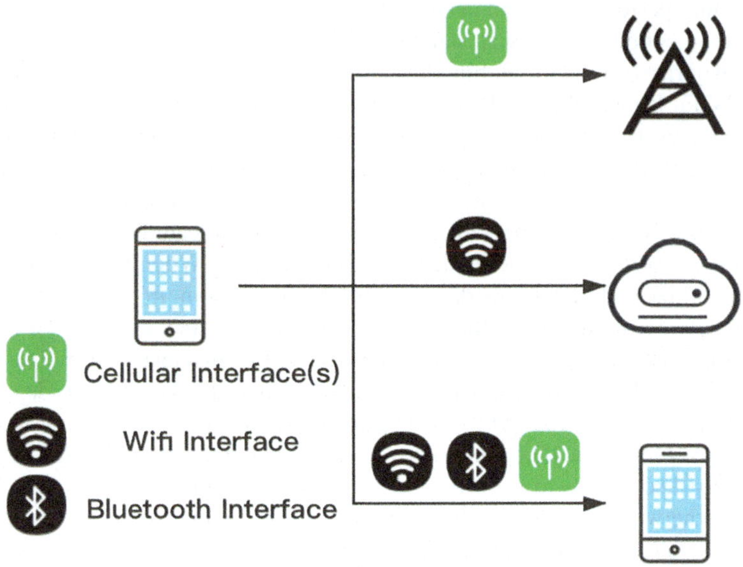

Fig. 4.1 Network connection with different interfaces for D2D

For instance, in the case of two candidate helper devices, one is viewing the video that the requester device is requesting, while the other is not. Because there is no need to use additional space to cache data, choosing the first helper device can significantly increase cache efficiency. Of the two candidate helper devices, the first device is only 20% charged while the second is 90%. It is more logical to select the second device because the establishment of a D2D connection will result in increased power consumption and low power warnings will cause users to feel negatively; for two candidate helper devices, the first wants to cache NBA game video to facilitate watching at any time, and the second is not interested in sports and wants to cache live game video. When a device requests an NBA game video, it is more reasonable to select the first to establish D2D connection. Considering the status information of candidate devices comprehensively, dynamic smart decision for D2D helper devices selection can ensure the video quality of requesting devices, improve the satisfaction of helper devices, and then improve the cooperative enthusiasm of cooperative devices, and promote the robustness of the whole video streaming transmission network based on D2D technology.

Moreover, the majority of modern mobile devices offer a variety of network interfaces, such as WiFi, Bluetooth, and other cellular network interfaces (Fig. 4.1). To meet diverse data transmission requirements across multiple network interfaces, multiple network connections can be established. Mobile cellular users generally use the cellular network interface to establish high-speed data transmission, as opposed to Bluetooth technology, which are both largely used for low-power data transfer. It is reasonable to use multiple network interfaces on a single device to connect to multiple D2D cooperation devices and implement multi-path parallel transmission to

Table 4.1 Key terms and descriptions

Terms	Descriptions
BS, ES	Base station, edge server
RD, HD	Requester device set, helper device set
$L^t_{i,x}$	Connection status between RD i and x at time t
$lc_{i,x}$, $lw2_{i,x}$, $lw5_{i,x}$, $lb_{i,x}$	Connection status through cellular, 2.4GHz WiFi, 5GHz WiFi and Bluetooth between RD i and x
$I_j(t)$	Video intention of j at time t
$m_x(t)$	Watching status of the device x at time t
$p_j(t)$	State of charge for HD j at time t
$c_j(t)$	State of cache for HD j at time t
$q_j(t)$	Location in the transport of HD j at time t
$l_j(t)$	Available network interfaces of HD j at time t
$V^t_{i,x}$	Available datarate between RD i and x at time t
$\mathcal{W}_{i,x}$	Bandwidth resource of the connection between RD i and x
SINR_i	Signal-interference-noise ratio
$h_{i,x}$	The channel gain of the Rayleigh fading between device i and x
$P_{i,x}$	Transmit power of device x
$d_{i,x}$	Euclidean distance between RD i and x
α	Path loss exponent
σ^2	Noise power
Θ	Datarate for video with the highest resolution
$B_i(t)$	Overall datarate of RD i with multiple connections
$P_{\text{DD}}(\text{link})$	Power consumption for data transmission from HD to RD with different connections and link $\in$ {Cellular, 2.4GHzWiFi, 5GHzWiFi, Bluetooth}
$\mathbf{P}$	Power level change times of HDs during video transmission
$\mathbf{C}$	Caching efficiency of HDs during video transmission
$\mathbf{I}$	The completion of individual intention of HDs
Ω	Satisfaction utility of HD for video transmission
$\mathbf{Q}_{\text{gap}}$	Video quality gap
$\mathbf{Q}_{\text{switch}}$	Switch times of video resolution
Ψ	Video quality utility of RD

fully take advantage of the advantages of D2D cooperation, increase the available bandwidth of the requesting device, and improve the quality of the video service. The Transmission Control Protocol (TCP) evolution known as multi-path TCP (MP-TCP) provides protocol capability for concurrent data delivery over several network connections. Existing approaches, which infrequently consider employing multiple network interfaces of mobile devices to undertake simultaneous data transfer at the same time, do not completely exploit the benefits of D2D transmission.

In the following part, the device selection for D2D-assisted multi-path video transmission network system model will be presented and the problem formulation will be described. Key terms and descriptions are presented in Table 4.1.

Suppose there is a scenario, that has a video transmission network with a base station (BS), a few mobile edge servers (ES), and plenty of mobile devices in a bus, train, or tram. Within the effective distance, a connection can be established between devices, between devices and ES, and between devices and BS to effectuate video transmission. Once a D2D connection has been made between the two devices, the device requesting video service is referred to as the requesting device (RD), and the device assisting in video transmission for the RD is referred to as the helper device (HD).

It assumes that each RD has one WiFi network interface, one Bluetooth network interface, and two cellular network interfaces that support 2.4 GHz and 5 GHz. When an RD seeks video resources, it will initially connect to the BS over the cellular interface for each request. The RD continues to receive data from the BS and asks for assistance from the ES on both sides of the road and the HDs nearby when the BS is unable to provide smooth video resources with high resolution for the RD, or when the available bandwidth between the RD and the BS is less than the original video datarate Θ. The WiFi interface with 2.4 GHz or 5 GHz can be used by the RD to connect to the ES for data transmission, and the Bluetooth, WiFi, or cellular interfaces can be used by the RD to connect to the HD for data transfer.

The link between an RD i and a device x at time t is represented by a binary variable $L_{i,x}^{t}$, which is defined as follows:

$$L_{i,x}^{t} = \begin{cases} 1, & \text{If the device } i \text{ is connected with } x \\ 0, & \text{otherwise} \end{cases} \tag{4.1}$$

Utilizing four comparable binary variables $\mathrm{lc}_{i,\ x}$, $\mathrm{lw2}_{i,\ x}$, $\mathrm{lw5}_{i,\ x}$, and $\mathrm{lb}_{i,\ x}$ to indicate the connection through Cellular, 2.4 GHz WiFi, 5 GHz WiFi, and Bluetooth, respectively, we take into account the fact that many connections could be made simultaneously for each RD i. It meets requirements for all video transmission devices x and RD i.

$$\forall x, \forall i, L_{i,x}^{t} = \mathrm{lc}_{i,x}^{t} + \mathrm{lw2}_{i,x}^{t} + \mathrm{lw5}_{i,x}^{t} + \mathrm{lb}_{i,x}^{t} \in \{0, 1\} \tag{4.2}$$

There are six parameters for each HD $I_j(t)$, $m_j(t)$, $p_j(t)$, $c_j(t)$, $q_j(t)$, $l_j(t)$, which indicates the intention to view a video, the status of watching it, the charge level, the cache state, the position in the transport, and the current network interfaces. $I_j(t) \in [0, N]$ represents the video's current goal of j at time t. $I_j(t) = 0$ displays no regard for any video. $m_j(t) \in [0, N]$ depicts the viewing state of j at that moment t. $m_j(t) = 0$ indicates that the device j is not watching the video. $l_j(t) \in \{\text{Cellular}, \text{2.4GHzWiFi}, \text{5GHzWiFi}, \text{Bluetooth}\}$.

According to Shannon Theory,

$$V_{i,x}^{t} = \mathcal{W}_{i,x} \log_2(1 + \mathrm{SINR}_i) \tag{4.3}$$

$\mathcal{W}_{i,x}$ is the connection's bandwidth resource, and various network connections have varied bandwidths. So that the signal-interference-noise ratio (SINR) is:

$$\mathrm{SINR}_i = \frac{h_{i,x} P_{i,x} d_{i,x}^{-\alpha}}{\sum\limits_{x \in \mathrm{HD},\, x' \neq x,\, L_{i,x'} = L_{i,x}} L_{i,x'} h_{i,x'} P_{i,x'} d_{i,x'}^{-\alpha} + \sigma^2} \tag{4.4}$$

Consequently, the total datarate of RD i is:

$$B_i(t) = \min\left\{ L_{i,\mathrm{BS}}^t V_{i,\mathrm{BS}}^t + L_{i,\mathrm{ES}}^t V_{i,\mathrm{ES}}^t + \sum_{j=1}^{M} L_{i,j}^t V_{i,j}^t, \Theta \right\} \tag{4.5}$$

Here, the function min{} can manage the datarate that is available to implement standard video playback and stop bandwidth resources from being wasted.

In order to increase HDs' satisfaction, it is important to identify the connections between devices that are most logical and can guarantee that the total available datarate of RD at any given time reaches the typical video datarate Θ as closely as feasible. The management of power usage, cache effectiveness, and the accomplishment of individual intention are all indicators of HD's pleasure.

For the control of power consumption **P**, empirically divide the electricity of the device into five levels: 1–5 corresponds to [0–10%], [10%–20%], [20%–50%], [50%–80%] and [80%–100%], respectively. Use $\Phi(p_j(t))$ to calculate the power level of HD j at time t:

$$\Phi(p_j(t)) = \begin{cases} 1, & 0 \leq p_j(t) < 10, \\ 2, & 10 \leq p_j(t) < 20, \\ 3, & 20 \leq p_j(t) < 50, \\ 4, & 50 \leq p_j(t) < 80, \\ 5, & 80 \leq p_j(t) \leq 100. \end{cases} \tag{4.6}$$

As a result, the following formula can be used to determine the overall power consumption controlling score:

$$\mathbf{P}(t) = \sum_{j=1}^{M} \sum_{i=1}^{N} L_{i,j}^t \left[\Phi\left(p_j(t) - P_{DD}(l_j(t))\right) - \Phi(p_j(t)) \right] \tag{4.7}$$

Determine if the video being seen by HD at the moment matches the video that RD needs to be delivered by HD to determine the caching efficiency **C**. If so, it indicates that HD does not require more storage space, and the caching efficiency is 1 in such case. If not, it is 0. Following is an expression for the overall caching efficiency for all HDs simultaneously at time t:

$$\mathbf{C}(t) = \sum_{j=1}^{M} \sum_{i=1}^{N} L_{i,j}^{t} V_{ij}^{t} \Delta t \left[m_j(t) \odot m_i(t) \right] \tag{4.8}$$

To fulfill each particular intention **I**, determine whether the video that HD needs and that RD is transmitting is the video that HD intends to cache. If so, then the fulfillment of HD's individual intention is 1, and if not, then it is 0. In addition, the total amount of individual intention for all HDs at a given time can be calculated as follows:

$$\mathbf{I}(t) = \sum_{j=1}^{M} \sum_{i=1}^{N} L_{i,j}^{t} \left[I_j(t) \odot m_i(t) \right] \tag{4.9}$$

When participating in video transmission at time t, the overall utility function of HD can be represented as follows:

$$\Omega(t) = \lambda_1 \mathbf{P}(t) + \lambda_2 \mathbf{C}(t) + \lambda_3 \mathbf{I}(t) \tag{4.10}$$

Where the reflection parameters for the level of constraint violation are λ_1, λ_2, and λ_3.

Enhancing video quality is the major goal to keep in mind because it is the main goal of D2D-assisted multi-path video transmission to increase the video quality of RDs. Quality gap and switch times are the two main video indicators to pay attention to. The disparity between the received video quality and the original video quality is represented by the quality gap. The following formula can be used to determine the quality gap of the video that RD received at the time t:

$$\mathbf{Q}_{\text{gap}}^{i}(t) = |B_i(t) - \Theta| \tag{4.11}$$

Whether the resolution switched of the video received by RD i at time t can be determined by:

$$\mathbf{Q}_{\text{switch}}^{i}(t) = \begin{cases} 1, & B_i(t) = B_i(t-1) \\ 0, & \text{otherwise} \end{cases} \tag{4.12}$$

The utility function for video quality in RDs at the time t can thus be written as follows:

$$\Psi(t) = \lambda_4 \sum_{i=1}^{N} \mathbf{Q}_{\text{gap}}^{i} + \lambda_5 \sum_{i=1}^{N} \mathbf{Q}_{\text{switch}}^{i} \tag{4.13}$$

The overall utility function at time t can be expressed as follows:

$$\max[\Omega(t) + \Psi(t)] \tag{4.14a}$$

s.t. (4.7), (4.8), (4.9), (4.10), (4.11), (4.12), (4.13)

$$\sum_{j=1}^{M} L_{i,j}^{t} + \left(V_{i,\mathrm{BS}}^{t} \odot 0\right) + \left(V_{i,\mathrm{ES}}^{t} \odot 0\right) \leq 4 \tag{4.14b}$$

$$\left(V_{i,\mathrm{BS}}^{t} \odot 0\right) + \sum_{j=1}^{M} lc_{i,j}^{t} \leq 2 \tag{4.14c}$$

$$\sum_{j=1}^{M} lb_{ij}^{t} \leq 1 \tag{4.14d}$$

$$\left(V_{i,\mathrm{ES}}^{t} \odot 0\right) + \sum_{j=1}^{M} \left(lw2_{i,j}^{t} + lw5_{i,j}^{t}\right) \leq 1 \tag{4.14e}$$

$$\sum_{j=1}^{M} lb_{i,j}^{t} + \sum_{j=1}^{M} lw2_{i,j}^{t} \leq 1 \tag{4.14f}$$

The restriction on the total number of network interfaces for an RD is given by Eq. (4.14b). The restriction on the number of cellular interfaces is given by Eq. (4.14c). The limit on the number of Bluetooth interfaces is given by Eq. (4.14d). The restrictions on the number of WiFi interfaces are given by Eq. (4.14e), whereas the restriction on connecting 2.4 GHz WiFi and Bluetooth is given by Eq. (4.14f).

It is necessary to dynamically choose the best HDs and the most logical connection mode at all times due to the unknown, rapidly changing network scene. Therefore, it is possible to acquire the connections mode set $\mathcal{L} = \left\{\operatorname{argmax}_{L_{i,j}^{t}} [\Omega(t) + \Psi(t)] | t \in [1, T]\right\}$. This inspires us to develop a dynamic device selection mechanism that carefully weighs the RD's current network condition as well as the distinctive data of HDs.

4.3 Dynamic Intelligent Decision in Multi-Path Parallel Transmission

A dynamic intelligent choice approach for helper device selection to support D2D-assisted multi-path video streaming network is proposed to address the aforementioned issues. Consider the edge computing environment's multi-path parallel video transmission network with D2D support in Fig. 4.2.

On a high-speed bus, there are lots of RDs and HDs, as seen in Fig. 4.2. The RD's network state is continually changing as a result of the bus' continuous movement.

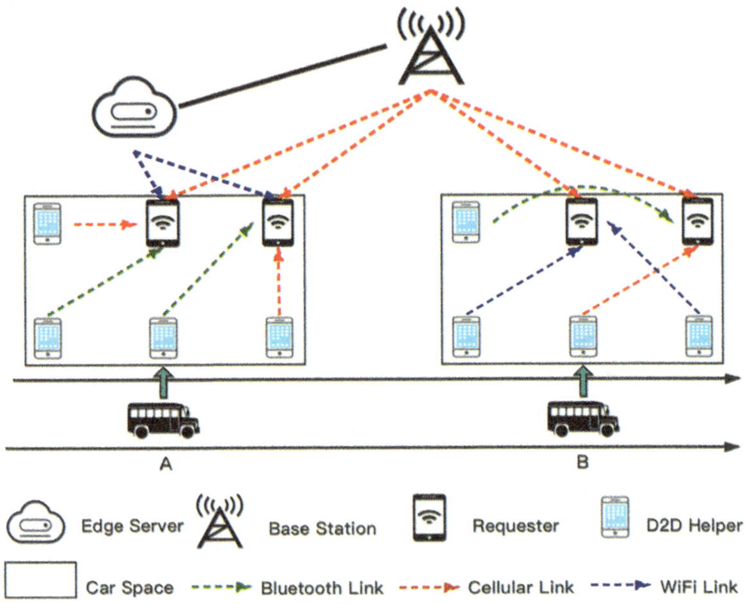

Fig. 4.2 The D2D-assisted multi-path video transmission network

For instance, at position A, an edge server controls transmission; however at location B, there is no edge server to assist. To increase HDs' enthusiasm for cooperation and to improve RDs' video quality, dynamically select different HDs based on the different network conditions. The component of this environment is as followed:

- **Video requesting device**: It can request video resources and includes a variety of network connections, including WiFi, Bluetooth, and cellular network interfaces. The MP-TCP protocol allows many network interfaces to establish connections with numerous devices and simultaneously receive video data;
- **D2D helper device**: It can create a D2D connection with the video requester over WiFi, Bluetooth, or cellular network interface to ease data transmission;
- **Base station**: It connects to the video requester via the cellular network interface and broadcasts streaming video to any cellular network user within range of the network;
- **Edge server**: Placed on both sides of the street, it transmits data to users inside the network's coverage and caches video, and it is connected to the device making the request through WiFi;
- **Bus**: It covers various cellular network base stations and edge servers while transporting numerous video-requesting devices and D2D helper devices at high speeds on the highway;
- **WiFi, Bluetooth, and Cellular connection**: Depending on the number of network interfaces in the video request device, various network connection modes

can be created with various devices (such as a cellular network base station, an edge server, and a device for two-way (D2D) communication).

The requesting devices and D2D helper devices are dynamic changing at different times because the network condition of mobile devices changes with the environment and other factors. Devices that can fulfill the requirements of normal video playback at the original datarate but do not watch video are known as D2D helper devices. The cooperative device communicates its status to the network, which also receives updates on its location on the vehicle, the network interface that is accessible, the purpose to cache videos, the status of viewers, the amount of power and cache space left, and other details. The requesting device carefully considers a number of factors and selects the most practical cooperation equipment to establish a D2D connection after receiving the collaboration device's status information to improve the video quality it receives, maximize its own cooperation satisfaction, and then improve the stability and dependability of the entire D2D cooperation network.

Algorithm 4.1: Watching State and Intention-aware HD Clustering Algorithm

Input: HD set $\mathcal{HD}$, RD set $\mathcal{RD}$**Output:** HD candidate sets S_1, S_2, S_3, S_4
for $j \in \mathcal{HD}$ **do**
if $\exists i \in \mathcal{RD}, \mathrm{m_j} = = m_i$ **then**
Add j to $S_1(i)$
else
if $\exists i \in \mathcal{RD}, \mathrm{l_j} = = m_i$ **then**
Add j to $S_2(i)$
else
if $l_j = = 0$ **then**
Add j to S_3
else
Add j to S_4
end if
end if
end if
end for

Algorithm 4.2: Power-Aware HD Selection Algorithm

Input: RD i, HD candidate sets S_1, S_2, S_3, S_4**Output:** HD candidate set C
for $S \in [S_1(i), S_2(i), S_3, S_4]$ **do**
for $j \in S$ **do**
if $c_j \geq V_{i,j} * \Delta t$ and $p_j \geq P_{DD}(l_j)$ **then**
if $\Phi(p_j) = = \Phi(p_j - P_{DD}(l_j))$ **then**
Add j to C_1
else
Add j to C_2
end if
end if
end for
Sort the HDs in C_1 and C_2 according to $V_{i,j}$ and obtained C_1^* and C_2^*

(continued)

12: Add $C_1^* + C_2^*$ to C
13: **end for**

As shown in Algorithm 4.1, add the HDs who are watching video v to the list of potential HD set S_1 for the RD who is requesting video v. According to Eq. (4.8), you should choose them to transmit the video they are watching first to enhance their caching effectiveness. Add the HDs that want to cache video v to the list of RD candidates for HD set S_2, and let them transmit first to optimize HD satisfaction, as shown by Eq. (4.9). Add HDs that don't intend to cache to the list of HD candidates S_3. Include the remaining HDs in the list of HD candidates S_4. Each RD has a different set of HD candidates, and all RDs share the same HD candidate sets. The four HD candidate sets are in this order $S_1 > S_2 > S_3 > S_4$. In order to increase HD pleasure, this can optimize the implementation of each HD's unique aim and caching effectiveness.

The HDs in the HD candidate set must then be sorted in accordance with power and cache limits as well as changes in power level. In Algorithm 4.2, a power-conscious HD selection algorithm is suggested. Go through each HD in turn in the four candidate HD sets that Algorithm 4.1 produced. Add HDs that do not modify their power levels after participating in D2D transmission into C_1 and add others into C_2 when the available cache space is larger than the video data to be cached and the available power is larger than the power required to create a D2D connection for data transmission.

C_1 has a greater priority than C_2 because power consumption control has a higher score according to Eq. (4.7). When a D2D connection is established to obtain C_1^* and C_2^*, HDs in C_1 and C_2 are then sorted by maximum bandwidth and added to the HD candidate set C. A prioritized HD candidate set C that thoroughly considers individual intention, cache efficiency, and power consumption controlling is eventually obtained after repeating the aforementioned actions on S_1, S_2, S_3, and S_4.

To get the best HDs, it is also necessary to maximize utility and consider the limitations of the network interfaces. Assume that each RD has two cellular network interfaces, one WiFi interface that can operate at 2.4 GHz or 5 GHz, and one Bluetooth interface. When broadcasting the video at the same time as 2.4 GHz WiFi and Bluetooth, there is significant signal interference that causes signal attenuation and data loss. The WiFi-based connection can only use one spectrum resource for data transfer at any given time since there is only one WiFi interface. In light of the aforementioned restrictions, Algorithm 4.3 is chosen to choose the ultimate HD set F.

Algorithm 4.3: Connection Method Selection Algorithm

Input: The HD candidate set C, RD i**Output:** The final HD set F
1: Calculate the required bandwidth $\mathcal{B} = \Theta - V_{i,ES} - V_{i,BS}$
2: $c = 2$
3: **if** $V_{i,\,BS} \neq 0$ **then**
4: $c = 1$
5: **end if**

(continued)

---

if $V_{i,\ ES} \neq 0$ **then**
  if the link between ES and i is 2.4GHz WiFi **then**
    Select the first c HD(s) from C which can generate Cellular-based D2D link and get the final HD set F
  else
    Select the first c HD(s) which can generate Cellular-based D2D link and 1st HD which can generate Bluetooth-based D2D link from C and get the final HD set F
  end if
else
    Select the first c HD(s) which can generate Cellular-based D2D link, 1st HD which can generate Bluetooth-based D2D link and 1st HD which can generate 5GHz WiFi-based D2D link from C and get the HD set f_1
    Select the first c HD(s) which can generate Cellular-based D2D link and 1st HD which can generate 5GHz WiFi-based D2D link from C and get the HD set f_2
    Select the first c HD(s) which can generate Cellular-based D2D link and 1st HD which can generate 2.4GHz WiFi-based D2D link from C and get the HD set f_3
    Calculate the utility of f_1, f_2 and f_3 according to Eq. (4.14a) and choose the most one as F
end if

For the utility function in Eq. (4.14a) to accurately reflect the real world, suitable weight values must be selected due to the various value ranges and evaluation criteria of the several attribute indexes. The technique for order preference by similarity to an ideal solution (TOPSIS) is used to settle this problem. TOPSIS is an effective method commonly used in multi-objective decision analysis. The distance between a number of evaluation objects and idealized targets is used by TOPSIS to rank the contenders to evaluate the relative merits of the many actual items. TOPSIS is a sorting approach that comes near to the optimum answer because each utility function simply needs to be monotonically growing (or decreasing). The ideal solution and the negative ideal solution are two key concepts in TOPSIS. When compared to the alternatives, the features of the putative best solution known as the ideal solution are the greatest, while those of the presumed worst solution known as the negative ideal solution are the worst. The fundamental idea is to compare the options to the ideal solution and the negative ideal. The best and most acceptable choice is the one that is farthest from the negative ideal solution yet closest to the ideal solution. The following are the TOPSIS steps:

1. For k candidate HD sets $f_1, f_2, \cdots, f_k$, each candidate has 5 attribute values x_1, x_2, x_3, x_4, x_5, which indicate **P**, **C**, **I**, $\mathbf{Q}_{gap}$, and $\mathbf{Q}_{switch}$.

$$F = \begin{bmatrix} x_{1,1} & \cdots & x_{1,5} \\ \vdots & \ddots & \vdots \\ x_{k,1} & \cdots & x_{k,5} \end{bmatrix} \tag{4.15}$$

   A candidate is represented by each row. Next, apply the following formula to each element of the matrix to normalize it:

$$y_{i,j} = \frac{x_{i,j}}{\sqrt{\sum_{i=1}^{k} x_{i,j}^2}} \tag{4.16}$$

where $i \in [1, k]$, $j \in [1, 5]$.

To obtain the weighted canonical decision matrix, multiply the norm value by the weight value $z_{i,j} = \alpha_j * y_{i,j}$ after empirically setting the starting weight coefficient for the five qualities α_1, α_2, α_3, α_4, α_5.

According to the weighted canonical value, the candidates' ideal solution $A^{+}$and negative ideal solution A^{-} are determined:

$$f^{+} = \{(\max\ z_{i,j} | j \in J_1), (\min\ z_{i,j} | j \in J_2) | i \in [1, k]\} = \{v_1^+, v_2^+, v_3^+, v_4^+, v_5^+\} \tag{4.17}$$

$$f^{-} = \{(\min\ z_{i,j} | j \in J_1), (\max\ z_{i,j} | j \in J_2) | i \in [1, k]\} = \{v_1^-, v_2^-, v_3^-, v_4^-, v_5^-\} \tag{4.18}$$

where J_1 indicates the ideal value on the i-th indication and is the collection of profit indicators. J_2 reflects the worst value on the i-th indication and is the set of loss indicators. The evaluation result is more beneficial the greater the profit indicator is, and the smaller the loss index, the more favorable the evaluation result is. It is detrimental to the outcomes of the evaluation, on the other hand.

The deviation degree of each candidate from the ideal solution S^{+} and the negative ideal solution S^{-} are computed based on Euclidean distance. The distance between the candidate and the ideal solution is while the distance between the applicant and the undesirable ideal solution is:

$$S^{+} = \sqrt{\sum_{j=1}^{5} \left(z_{i,j} - v_j^+\right)^2} \tag{4.19}$$

$$S^{-} = \sqrt{\sum_{j=1}^{5} \left(z_{i,j} - v_j^-\right)^2} \tag{4.20}$$

2. The calculation of the distance between each contender and the ideal solution is:

$$C_i = \frac{S_i^-}{S_i^+ + S_i^-} \tag{4.21}$$

3. The choices are ordered by how closely they are related, and the best option is chosen.

The DS-DAMP method is illustrated in Algorithm 4.4. For each time slot, Algorithm 4.1 is used to obtain the HD candidate sets $S1$, $S2$, $S3$, and $S4$ that take individual intentions and watching status into consideration. Algorithm 4.2 is used to obtain the HD candidate set that takes power and cache constraints and changes in power level into consideration. Algorithm 4.3 is then used to obtain the final HD set that satisfies network interface constraints and maximizes utility.

Algorithm 4.4: DS-DAMP

Input: Input: RD set $\mathcal{RD}$, HD set $\mathcal{HD}$
1: **while** $\mathcal{RD}$ is not null **do**
2: Process Algorithm 4.1 and get the HD candidate sets S_1, S_2, S_3, S_4 for each RD
3: **for** $i \in \mathcal{RD}$ **do**
4: Process Algorithm 4.2 and get the HD candidate sets C for RD_i
5: Process Algorithm 4.3 and get the final HD set F
6: **end for**
7: **end while**

4.4 Intelligent Multi-Access Device Cooperation in Multi-Path Network

Consider an edge computing network with a multi-path D2D-assisted video transmission system. Assume that N + M D2D-supporting devices are deployed in a bus with a length of 20 meters in accordance with the homogeneous Poisson Point Process, and that BSs and ESs are deployed at a fixed location in a round area with a radius of 1000 meters. Table 4.2 shows a list of the simulation's parameters and values.

Utilized are actual measurements and evaluations of the WiFi and 4G cellular networks. 1000 independent trials were conducted using various combinations of N and M, and the performance was gauged by averaging the results. The numbers of

Table 4.2 Simulation parameters and values

Simulation parameters	Values
The number of RD and HD N, M	1–9, 1–30
The original video datarate Θ	20 Mb/s
The time slot Δt	1 s
The path loss exponent α	2.6
α_1, α_2, α_3, α_4, α_5	−1, 0.2, 1, −0.1, −1
The radius of the video transmission area	1000 m
The location and the effective distance of the BS	500, 500
The location and the effective distance of the ES	[100, 500, 900], 100
Maximum transmit power of BS, cellular user, WiFi user, Bluetooth user	43, 30, 23, 23 dbm
Channel bandwidth of cellular, 2.4 GHz WiFi, 5 GHz WiFi, Bluetooth	20, 20, 40, 1 MHz
Noise spectral density	−174 dBm/Hz

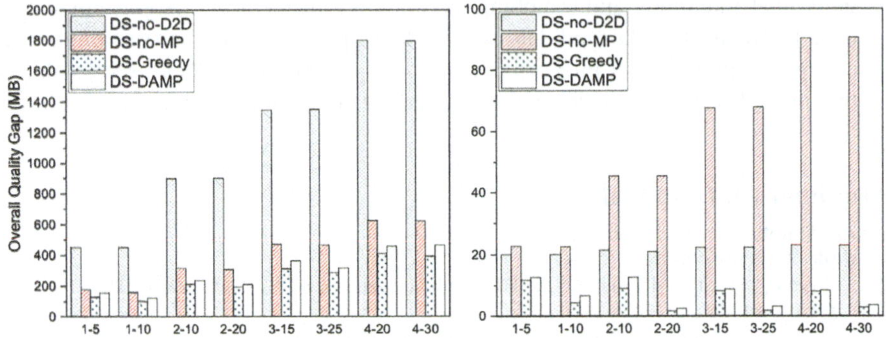

(a) Overall quality gap of the video with the highest resolution.

(b) Switching times during the video playback in RDs.

Fig. 4.3 Video quality of RDs. (**a**) Overall quality gap of the video with the highest resolution. (**b**) Switching times during the video playback in RDs)

RD and HD were, respectively [1, 5], [1, 10], [2, 10], [2, 20], [3, 15], [3, 25], [4, 20], [4, 30].

Three video transmission methods are chosen for comparison to more fairly assess the method's performance:

- **DS-no-D2D**: Only RD's Cellular transmission is used because there are no HDs or ES to aid in the transmission.
- **DS-no-MP**: Although numerous network interfaces are not required for multi-path parallel transmission, D2D is employed to aid transmission. There is only ever one network interface.
- **DS-Greedy**: D2D will be used to select the highest available bandwidth from among many HDs at once. The HDs' status information and intentions, however, are not taken into account.

First, various approaches are used to test the video quality that RDs receive. Figures 4.3a and b show, respectively, the overall quality gap (OQG) and switch times (ST). Because both will lower video quality, reducing OQG and ST can considerably improve the overall video quality of RDs. Next, the video quality gap and resolution changeover durations are combined to provide the RDs' video quality utility, which is shown in Fig. 4.4. To facilitate observation, a 50 percent positive offset was added to each set of data. Without D2D transmission, the OQG of DS-no-D2D is extremely high, and D2D transmission can greatly enhance the video quality that RDs receive. With DS-Greedy, the best video quality in terms of OQG and ST may be attained. This is primarily due to DS-Greedy's disregard for the HDs' status information and intentions. The RD has the finest video quality, but at a price that greatly lowers HD satisfaction, which could dampen HD cooperation enthusiasm and jeopardize the stability of the D2D network.

The satisfaction of HDs in comparison to DS-Greedy and DS-no-MP is next evaluated. Avoiding power level variations is a crucial consideration for DS-DAMP

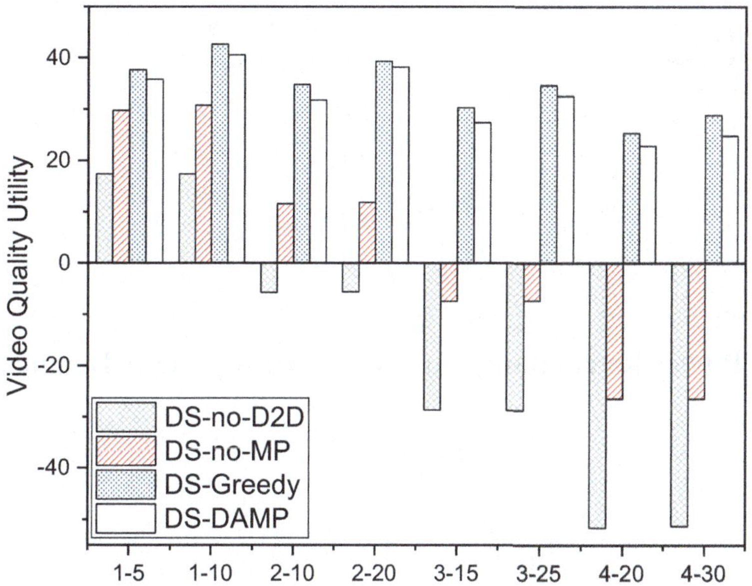

Fig. 4.4 RDs' video quality utility

as these changes will have a negative influence on HD users' enjoyment and reduce their motivation for cooperation. Figure 4.5a shows the variations in the power level of HDs for different numbers of HDs and RDs. The DS-DAMP approach consumes more power in HDs because it creates many D2D connections as opposed to DS-no-MP's single D2D connection at a time. In contrast, in terms of the quantity of power level fluctuations, surpassed DS-Greedy and nearly matched DS-no-MP. Because HD collaborative transmission of the video being watched can use the cache space rather than utilizing additional space to store temporary video data, the improved cache efficiency can effectively increase cooperative pleasure. As can be seen in Fig. 4.5b, DS-DAMP performs noticeably better than the other two approaches in terms of cache consumption. Since each HD has a unique desire to collaborate, aiming to satisfy each HD's desire can directly increase cooperation satisfaction. Intention completion for the HDs is depicted in Fig. 4.5c. Furthermore, it is obvious that DS-DAMP is better than the other two. Combine power level fluctuations, cache efficiency, and the success of individual intentions to estimate the usefulness of the final HDs. Figure 4.6 illustrates how DS-DAMP greatly outperforms the other two techniques.

The utility of DS-DAMP for the stability and resilience of video streaming networks is demonstrated in Fig. 4.7, which shows how it has reached the highest utility value in various networks with various numbers of RDs and HDs.

The number of HDs and RDs is constantly changed to further investigate the effects of various RD and HD counts on the network utility of video streaming transmission networks. Increase the number of RD N after fixing the total number of

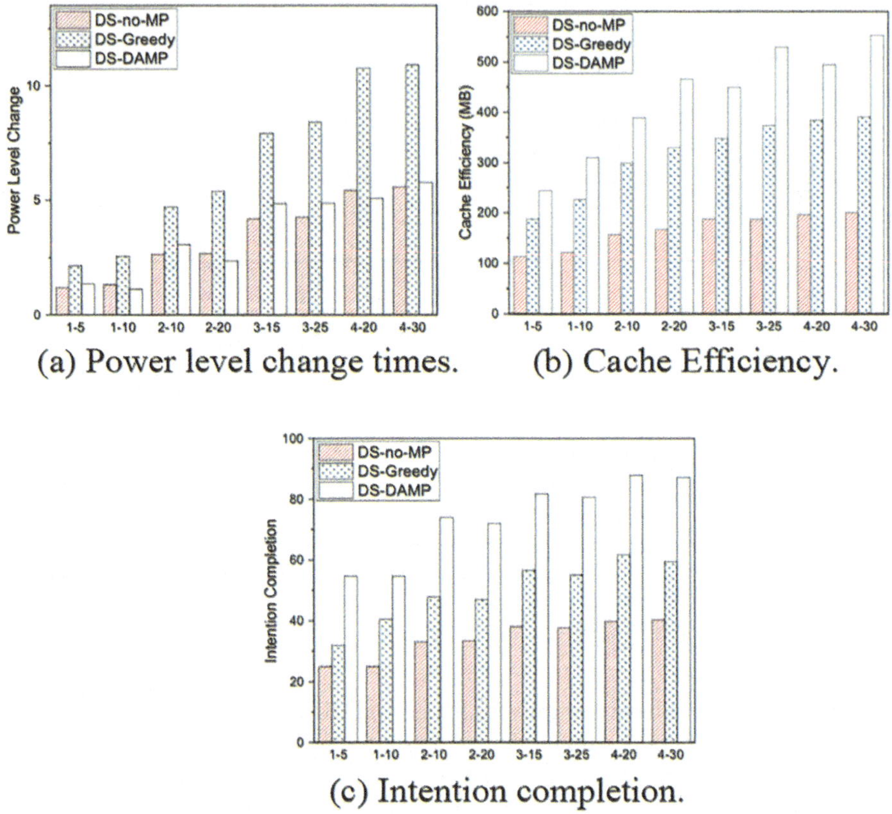

(a) Power level change times. (b) Cache Efficiency.

(c) Intention completion.

Fig. 4.5 Cooperation satisfaction of HDs. (**a**) Power level change times. (**b**) Cache Efficiency. (**c**) Intention completion

RDs and HDs, which is $M + N = 10$, as shown in Fig. 4.8. The network utility can be observed to be continually falling, especially as the number of RDs approaches or even exceeds the number of HDs. Then, as shown in Fig. 4.9, fix the number of RDs N and alter the number of HDs to observe the modification in network utility. Network utility exhibits a developing trend for the same N with a growing number of HD. This is done so that when creating D2D connections, RDs have a large enough pool of candidates to choose from. More HDs translate to more RD applicants. In order to construct the optimum D2D connection, RD can completely account for a number of factors. As a result, there needs to be enough HD participation for the D2D-based video streaming network to function at its best. In order to boost HDs' passion for their cooperation and to increase HD engagement, DS-DAMP completely considers HDs' satisfaction with their cooperation.

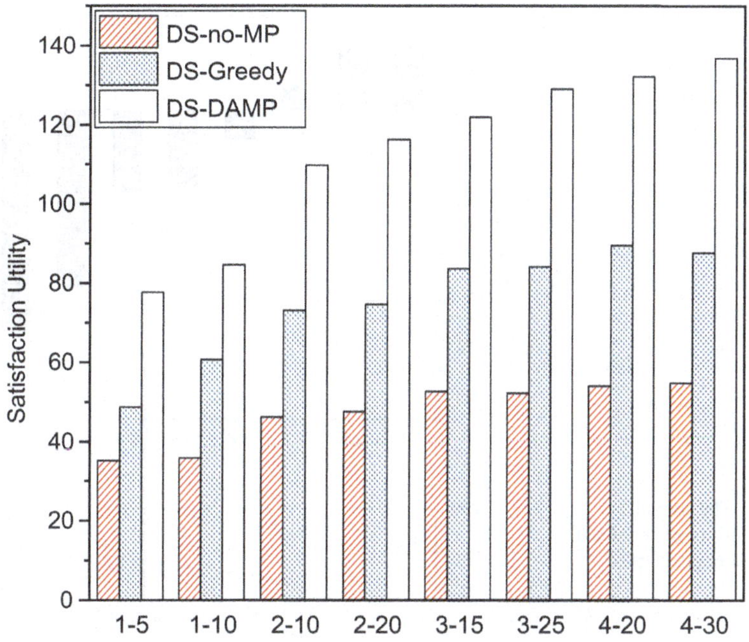

Fig. 4.6 HD's satisfaction utility for cooperation in video streaming transmission

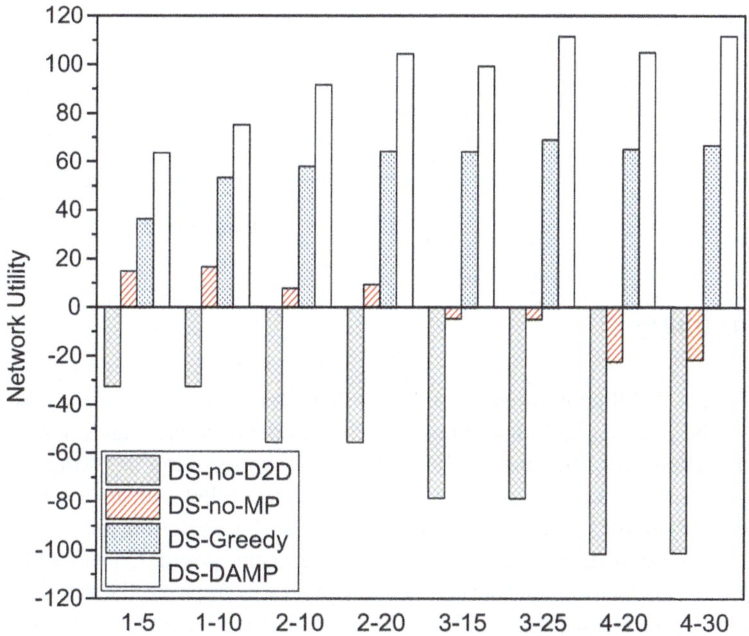

Fig. 4.7 Network utility of D2D-assisted multi-path video streaming transmission network

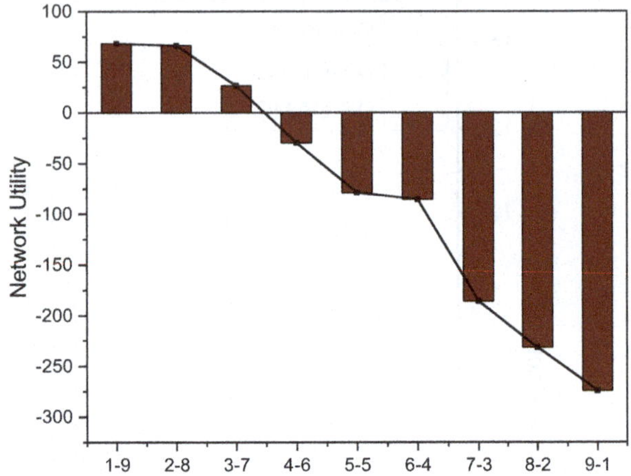

Fig. 4.8 The overall network utility changes with a fixed total number of D2D-supported devices

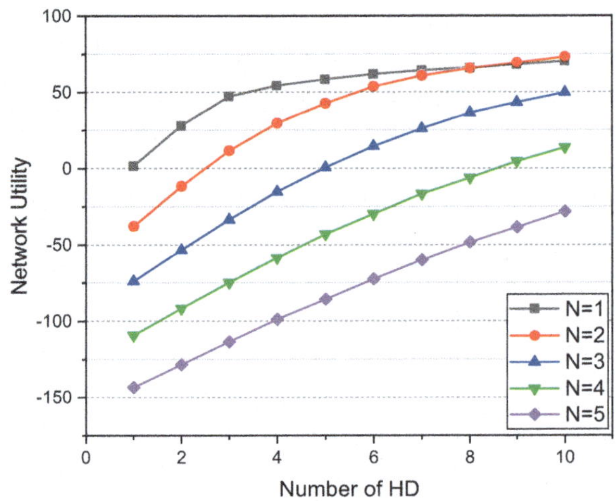

Fig. 4.9 The overall network utility changes with a fixed number of RDs

4.5 Summary

In this chapter, intelligent cooperation technology for multi-access device is introduced and the benefits and challenges of D2D cooperation are highlighted. Algorithms and criteria for selecting the best devices for cooperation are described, and dynamic decision-making algorithms for multi-path parallel transmission are also discussed. A dynamic smart decision method for device selection in D2D-assisted multi-path video transmission (DS-DAMP) network is proposed. Consider the status and intention to select the appropriate helper devices to improve the satisfaction of HDs and the video quality of RDs. In addition, the importance of intelligent cooperation for improving multi-access device communication is emphasized.

References

1. Josilo S, Dán G (2019) Selfish decentralized computation offloading for mobile cloud computing in dense wireless networks. IEEE Trans Mob Comput 18(1):207–220
2. Josilo S, Dán G (2019) Wireless and computing resource allocation for selfish computation offloading in edge computing. In: 2019 IEEE Conference on Computer Communications, pp 2467–2475
3. Sun H, Wang J, Peng H, Song L, Qin M (2021) Delay constraint energy effient cooperative offloading in MEC for IoT. In: Collaborative Computing: Networking, Applications and Worksharing, pp 671–685
4. Liang Y, Ge J, Zhang S, Wu J, Tang Z, Luo B (2019) A utility-based optimization framework for edge service entity caching. IEEE Trans Parallel Distributed Syst 30(11):2384–2395
5. Malik R, Vu M (2020) Energy-effient computation offloading in delay-constrained massive MIMO enabled edge network using data partitioning. IEEE Trans Wirel Commun 19(10): 6977–6991
6. Yang Y, Long C, Wu J, Peng S, Li B (2021) D2D-enabled mobile-edge computation offloading for multi-user IoT network. IEEE Internet Things J 8:12490–12504
7. Zhang X, Zhu Q (2019) D2D offloading for statistical QoS provisionings over 5G multimedia mobile wireless networks. In: IEEE INFOCOM 2019 - IEEE Conference on Computer Communications, pp 82–90
8. Saleem U, Liu Y, Jangsher S, Tao X, Li Y (2020) Latency minimization for D2D- enabled partial computation offloading in mobile edge computing. IEEE Trans Veh Technol 69(4): 4472–4486
9. Ko H, Pack S (2020) Distributed device-to-device offloading system: design and performance optimization. IEEE Trans Mob Comput 20:2949–2960
10. Sun M, Xu X, Tao X, Zhang P, Leung VCM (2020) NOMA-based D2D-enabled traffic offloading for 5G and beyond networks employing licensed and unlicensed access. IEEE Trans Wirel Commun 19:4109–4124
11. Pan Y, Pan C, Yang Z, Chen M, Wang J (2020) A caching strategy towards maximal D2D assisted offloading gain. IEEE Trans Mob Comput 19(11):2489–2504
12. Borgohain P, Choudhury H (2023) A lightweight D2D authentication protocol for relay coverage scenario in 5G mobile network. Comput Netw 225:109679
13. Ibrahim R, Assaad M, Sayrac B (2023) A dynamic policy for selecting D2D mobile relays. Comput Netw 221:109505
14. Feng B, Zhang C, Liu J, Fang Y (2019) D2D communications-assisted traffic offloading in integrated cellular-WiFi networks. IEEE Internet Things J 6(5):8670–8680
15. Peng J, Qiu H, Cai J, Xu W, Wang J (2021) D2D-assisted multi-user cooperative partial offloading, transmission scheduling and computation allocating for MEC. IEEE Trans Wirel Commun 20:4858–4873

Chapter 5
Deep Optimization for Storage and Computation Over Industrial Internet

Abstract The Industrial Internet has revolutionized the way businesses operate by integrating Internet technology with industrial processes. With the advent of the Industrial Internet, the integration of data and automation has become possible, creating new opportunities for businesses to optimize their operations and improve their efficiency. Intelligent technologies, including deep learning, attentive model compression, computation sharing, and deep optimization, are driving a transformation in the Industrial Internet. These technologies empower businesses with data-driven insights, enabling them to make informed decisions and streamline their workflows.

Deep learning has played a significant role in optimizing storage and computational resources. With the development of deep learning techniques such as neural network compression and quantization, deep learning models can be deployed in industrial equipment in a smaller and more efficient manner without sacrificing accuracy. This allows for more efficient use of storage space and computational resources, which is crucial in industrial applications where resource efficiency is essential. In addition, techniques such as transfer learning allow for the reuse of pre-trained models, further reducing the need for extensive computational resources. Overall, deep learning has helped to make intelligent industrial applications more efficient and cost-effective by optimizing the use of storage and computational resources.

The Industrial Internet has revolutionized the way businesses operate by integrating Internet technology with industrial processes. With the advent of the Industrial Internet, the integration of data and automation has become possible, creating new opportunities for businesses to optimize their operations and improve their efficiency. Intelligent technologies, including deep learning, attentive model compression, computation sharing, and deep optimization, are driving a transformation in the Industrial Internet. These technologies empower businesses with data-driven insights, enabling them to make informed decisions and streamline their workflows.

Deep learning has played a significant role in optimizing storage and computational resources. With the development of deep learning techniques such as neural network compression and quantization, deep learning models can be deployed in

W. Dou et al., *Intelligent Industrial Internet Systems*,
https://doi.org/10.1007/978-981-99-5732-3_5

industrial equipment in a smaller and more efficient manner without sacrificing accuracy. This allows for more efficient use of storage space and computational resources, which is crucial in industrial applications where resource efficiency is essential. In addition, techniques such as transfer learning allow for the reuse of pre-trained models, further reducing the need for extensive computational resources. Overall, deep learning has helped to make intelligent industrial applications more efficient and cost-effective by optimizing the use of storage and computational resources.

5.1 Deep Learning-Empowered Industrial Internet

Deep learning is a potent tool for processing and analyzing vast amounts of data, rendering it a prime technology for many applications, such as computer vision (CV), natural language processing (NLP), audio processing, and autonomous vehicles, where the key is to improve the performance of deep learning models [1–5]. In contrast to these classical applications, the resource consumption of deep learning models is more important than their performance due to the severe limitations on the system capacity (e.g., CPU, GPU, memory, etc.) of industrial devices in the Industrial Internet [3, 6, 7].

This section explores the several classical ways in which deep learning is leveraged to empower the Industrial Internet, along with the critical challenges that businesses should surmount to effectively implement intelligent applications across the Industrial Internet.

Specifically, we first present methods to empower the Industrial Internet, such as predictive maintenance, quality control, and supply chain optimization.

Predictive maintenance. Predictive maintenance is one of the most significant applications of deep learning in the Industrial Internet. Traditional maintenance practices are often reactive, meaning that maintenance is only performed when a machine or equipment fails. Predictive maintenance, on the other hand, is proactive, meaning that maintenance is performed before a machine or equipment fails, thereby preventing downtime and reducing maintenance costs.

Deep learning algorithms can be used to analyze data from sensors and other sources to identify patterns that indicate when equipment is likely to fail. By analyzing data on factors such as temperature, vibration, and pressure, deep learning algorithms can predict when equipment is likely to fail and alert maintenance teams to perform preventative maintenance.

For example, in the aviation industry, predictive maintenance is used to analyze data from sensors on aircraft engines. Deep learning algorithms allow maintenance teams to replace components before they fail by analyzing the collected data, reducing the risk of in-flight failures and improving aircraft safety.

Similarly, in the manufacturing industry, predictive maintenance is used to analyze data from sensors on production equipment. By analyzing data collected by various sensors, deep learning algorithms can predict when equipment is likely to fail and alert maintenance teams to perform preventative maintenance. This can reduce downtime and maintenance costs, improving the overall efficiency of the manufacturing process.

Quality control. Quality control, which refers to the measures and methods taken in the production process to ensure that the product or service meets the expected quality standards, is another significant application of deep learning in the Industrial Internet. Traditional quality control practices involve manual inspection of products, which can be time-consuming and prone to error. Deep learning algorithms can be used to automate quality control by analyzing images or sensor data to identify defects or anomalies.

For example, in the automotive industry, deep learning algorithms are used to analyze images of car body panels to identify defects such as dents or scratches. This allows manufacturers to identify and correct defects early in the production process, reducing waste and improving product quality.

In the food and beverage industry, deep learning algorithms are used to analyze sensor data to identify contaminants or other quality issues. By analyzing data on factors such as temperature, pressure, and PH, deep learning algorithms can identify potential quality issues before products are shipped to customers, reducing the risk of product recalls and improving customer satisfaction.

Supply chain optimization. Deep learning algorithms can also be used to optimize supply chain operations in the Industrial Internet. By analyzing data on factors such as demand, inventory levels, and shipping times, deep learning algorithms can predict when and where inventory needs to be replenished and can optimize shipping routes to reduce costs and improve delivery times.

For example, in the logistics industry, deep learning algorithms are used to optimize shipping routes based on factors such as traffic conditions, weather, and delivery deadlines. This can reduce shipping costs and improve delivery times, improving customer satisfaction and reducing the risk of product spoilage.

In the retail industry, deep learning algorithms are used to predict demand for products and optimize inventory levels. By analyzing data on factors such as sales history, weather, and promotions, deep learning algorithms can predict when and where products will be in demand and can adjust inventory levels accordingly. This can reduce inventory costs and improve customer satisfaction by ensuring that products are always available when customers want to purchase them.

Despite the benefits of deep learning in the Industrial Internet, there are also several challenges that businesses must overcome to successfully implement intelligent applications, including: (a) **Resource consumption:** Deep learning algorithms require significant computational resources to train and run, which can be a challenge for businesses with limited computing infrastructure. In addition, deep learning algorithms can generate large amounts of data, which can be difficult to store and manage; (b) **Interpretability:** Deep learning algorithms can be complex and difficult to understand, making it challenging for businesses to interpret their results and make decisions based on them. In addition, deep learning algorithms can be prone to overfitting, where the algorithm is too closely tuned to the training data and is unable to generalize to new data; (c) **Data quality and availability:** Deep learning algorithms rely on high-quality data to generate accurate predictions and insights. However, high data quality requirement is a challenge in the Industrial Internet, where data can be noisy, incomplete, or inconsistent. In addition, data availability can also be a challenge, as some sensors are not able to transmit data in real-time or may not be accessible in remote locations.

To overcome the resource consumption problem, enterprises can use cloud computing services, which provide on-demand access to computing resources and storage. Furthermore, businesses can also use edge computing and fog computing, which enable data to be processed and analyzed closer to the source, reducing the need for data to be transmitted over long distances [8, 9]. To overcome the data quality and availability problem, businesses should invest in data quality and data management processes that ensure that data is accurate, complete, and consistent. In addition, businesses should also invest in technologies such as edge computing and fog computing, which enable data to be processed and analyzed closer to the source, avoiding the need to transmit data over long distances [10, 11]. To overcome the algorithm complexity problem, businesses can invest in training programs and resources that enable employees to understand and interpret the results generated by deep learning algorithms [12, 13]. In addition, enterprises can also use techniques such as model regularization to reduce the risk of overfitting and ensure that algorithms generalize well to new data.

In conclusion, deep learning is a powerful technology that is transforming the Industrial Internet by providing businesses with data-driven insights that can be used to optimize their operations and improve their efficiency. However, businesses need to overcome several challenges to successfully implement deep learning applications, including resource consumption, data quality and availability, and algorithm complexity. By investing in data quality and management processes, cloud computing services, and training programs, businesses can effectively unlock the potential of deep learning in the Industrial Internet.

5.2 Attentive Model Compression for Deep Learning Model

Deep learning has been widely used in various fields, such as computer vision, natural language processing, and speech recognition. However, deep neural network (DNN) models are usually large and computationally expensive, which limits their deployment on resource-constrained devices. Attentive model compression is an effective way to reduce the model size and computational complexity while maintaining high accuracy [14–16]. In this section, we will discuss attentive model compression for deep learning models and its application in industrial applications. Specifically, we will first introduce the basic concepts of model compression and how it reduces the model size while maintaining satisfying model accuracy. Then, we will present the CroApp framework [17], in which the attentive model compression is integrated for saving storage and computational resources caused by training deep learning model.

Model compression is a technique that reduces the size of deep learning models by removing redundant weights and structures while maintaining their accuracy [18–20]. There are three main approaches to model compression: pruning, quantization, and knowledge distillation.

Pruning is a widely used technique for reducing the storage and computational costs of deep neural networks [15]. It involves removing redundant weights and structures from the model, which can significantly reduce the model size and accelerate the inference process. There are two main types of pruning methods: structured pruning and unstructured pruning. Structured pruning is a popular pruning method that removes entire channels or filters in convolutional neural network (CNN) based on their importance scores. The importance score is usually determined by the magnitude of the weight or the activation. Structured pruning can achieve significant compression rates with little or no loss in accuracy. Moreover, it can be applied in a layer-wise or group-wise manner, which can further improve the compression rate. On the other hand, unstructured pruning removes individual weights from the model without affecting the overall structure of the network. This type of pruning can achieve higher compression rates than structured pruning, but it requires more computation and can lead to irregular structures that are difficult to optimize.

Quantization is a technique that aims to reduce the memory footprint and computational complexity of deep learning models [21]. By reducing the precision of the weights and activations from 32 bits to 8 bits or even lower, quantization can significantly reduce the memory and computation requirements of a model without sacrificing its accuracy. Quantization-aware training is a popular method that allows deep learning models to be trained directly with low-precision weights and activations, instead of the full precision values typically used in training. By using low-precision values from the start, the model can be trained to work optimally with them, resulting in a more accurate and efficient model. In addition, quantization-aware training can improve the robustness of the model to quantization errors, which can occur when the weights and activations are rounded to low-precision values.

Knowledge distillation is a technique that has been developed to address the challenge of deploying large, complex models in resource-constrained environments [22]. Knowledge distillation involves training a smaller model to learn from a larger, more complex model, effectively compressing its knowledge and improving its performance. This method has the potential to significantly reduce the size and computational requirements of models while still maintaining high accuracy. In knowledge distillation, the large, accurate model is referred to as the teacher model, while the smaller, more efficient model is known as the student model. The student model is trained to mimic the output of the teacher model by minimizing the difference between the two outputs. By doing so, the student model can learn from the teacher model and reproduce its predictions with a high degree of accuracy.

CroApp is a novel resource optimization framework that is used to reduce both the computational and storage resource consumption for industrial equipment over the Industrial Internet. The CroApp framework comprises two optimization

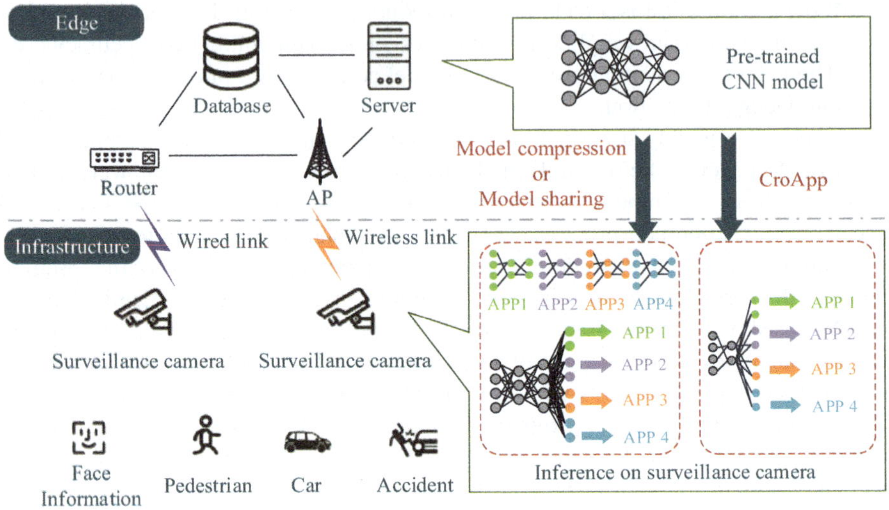

Fig. 5.1 An intelligent residential district with smart surveillance cameras

methods: an "inner-model" optimization method and an "inter-model" optimization method. Specifically, attentive model compression is explored as the "inner-model" optimization method, whereas computation sharing is explored as the "inter-model" optimization method. An example of an intelligent residential district with smart surveillance cameras is depicted in Fig. 5.1, in which the edge computing paradigm is enabled to allow data to be processed and analyzed closer to the source and avoid the need to transmit data over long distances. The router and access point (AP) are responsible for data transmission, and the database is responsible for storing data. In this scenario, the administrator of the intelligent residential district requires four CNN-based applications with various functions, including object detection, object classification, accident detection, and face recognition.

Although existing model compression approaches have been proposed to reduce the resource occupation of models by generating small yet efficient models, they inevitably incur redundant resource consumption. For example, in the intelligent residential district scenario, each surveillance camera requires the same CNN model (e.g., ResNet) for performing the tasks of all four applications. Model compression approaches compress the CNN model on the edge server for each of the applications and send the four identical compressed models to each of the surveillance cameras, resulting in redundant resource consumption. Pioneer computation-sharing approaches focus on efficiently adjusting the shareable resource occupation across applications to maintain satisfying application performance. However, these approaches ignore the original resource consumption within the model. For example, in the intelligent residential district scenario, computation-sharing approaches generate a single model for all four applications, without any reduction in the volume of the single model.

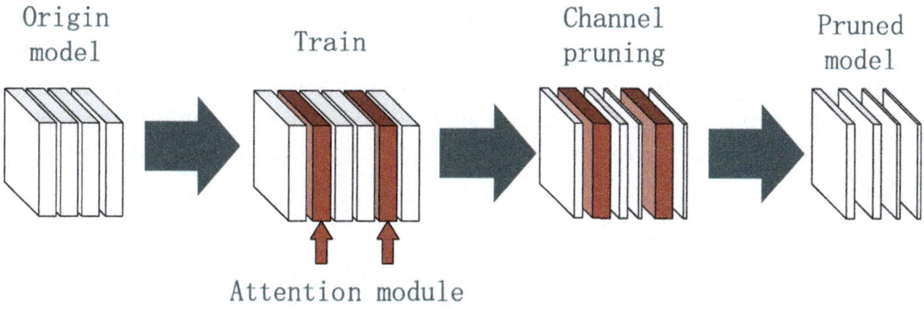

Fig. 5.2 Inner-model optimization with attention-based channel pruning

In contrast to the existing approaches, CroApp offers a novel solution by generating only a single compact model for all four applications, which is designed to share the compressed CNN model for inference. This approach offers superior resource optimization performance compared to state-of-the-art methods, which either incur redundant resource consumption or fail to reduce the volume of the single model. By sharing the same model across all four applications, CroApp can achieve significant reductions in resource consumption without sacrificing the performance of individual applications.

In the remainder of this section, we will provide a detailed introduction to the "inner-model" optimization method, while the "inter-model" optimization method will be discussed in the subsequent section.

We provide a detailed explanation of the "inner-model" optimization approach used in CroApp, which is depicted in Fig. 5.2. The approach involves inserting attention modules behind the layers that need to be pruned (i.e., target layers), obtaining importance scores during the training process, pruning the least important channels, and removing the inserted attention module to obtain the final pruned model.

Attention module design: The attention module in CroApp is designed with a spatial attention submodule and a channel attention submodule, arranged in a sequential "spatial-channel" fashion. The spatial attention submodule utilizes a grouping strategy [23] to reduce the computation overhead and obtain specific semantic responses. To obtain local features, the c channels are divided into g groups, where the height and width of each channel are h and w, respectively, and g is a pre-defined hyperparameter. The local feature vector $\boldsymbol{F}_{\text{loc}}$ is then calculated for each group as follows:

$$\boldsymbol{F}_{\text{loc}} = \{\boldsymbol{f}_1, \ldots, \boldsymbol{f}_n\}, \boldsymbol{f}_i \in \mathbb{R}^{\frac{c}{g}}, n = h \times w, \tag{5.1}$$

where $\boldsymbol{f}_i$ represents the local feature at every position. For global features, average-pooling and max-pooling are used to capture the global feature. Via the concatenate operation, the global feature $\boldsymbol{F}_{\text{glo}}$ can be calculated as follows:

$$\boldsymbol{F}_{\text{glo}} = \text{con}\left(P_{\text{avg}}(\boldsymbol{F}_{\text{loc}}), P_{\text{max}}(\boldsymbol{F}_{\text{loc}})\right), \tag{5.2}$$

where the con represents the concatenate operation, the P_{avg} represents the average-pooling operation, and the P_{max} represents the max-pooling operation. The importance coefficient of each position is then calculated as $S_i = \boldsymbol{F}_{\text{glo}} \cdot \boldsymbol{F}_{\text{loc}}$ and normalized as follows:

$$\widehat{S_i} = \frac{S_i - \mu_S}{\sigma_S + \varepsilon}, \mu_S = \frac{1}{n}\sum_j^n S_j, \sigma_S = \sqrt{\frac{1}{n}\sum_j^n (S_j - \mu_S)^2} \tag{5.3}$$

where the $\widehat{S_i}$ is the normalized importance coefficient, the ε is a constant added for numerical stability, the μ_S and σ_S are the mean and the standard deviation, respectively. The spatial attention is then calculated as $\Gamma\text{S} = \text{sig}\left(\widehat{S_i}\right)$, where the *sig* represents sigmoid function.

Similar to spatial attention, channel-wise features are extracted using average-pooling and max-pooling operations, and the channel attention submodule is calculated as follows:

$$\Gamma = \text{sig}\left(\text{GN}\left(P_{\text{avg}}(\Gamma\text{S})\right) + \text{GN}(P_{\text{max}}(\Gamma\text{S}))\right), \tag{5.4}$$

where GN represents the group normalization. The outputs of the channel attention submodule are the final outputs of the attention module.

Channel pruning: Due to the different input data generates different attention results [15], we first decrease the bias caused by input data. Specifically, importance score vector I is calculated as follows:

$$I = \frac{\sum_k^N \Gamma^k}{N} \tag{5.5}$$

where N represents the number of input data, and Γ^k represents the importance score vector of the data k. CroApp then prunes the channels with the least importance score using a pre-defined pruning ratio based on the importance score vector I. After pruning, the attention module is removed to prevent unnecessary costs. The detailed algorithm is shown in Algorithm 5.1.

Algorithm 5.1: Inner-Model Optimization Algorithm

Require: A CNN model $\mathcal{M}_{base}$ with parameters $\boldsymbol{\theta}_{base}$, attention module with parameter $\boldsymbol{\theta}_{att}$, N training batches, pre-defined prune ratio η.
Ensure: The compressed CNN model $\mathcal{M}_{pru}$ with fewer parameters $\boldsymbol{\theta}_{pru}$
1: Insert the attention module: $\boldsymbol{\theta}_{base} \leftarrow \boldsymbol{\theta}_{base} + \boldsymbol{\theta}_{att}$
2: **for** each batch k in N **do**
3: Train the batch of data and obtain channel attention Γ
4: **end for**

(continued)

5: Calculate importance score I
6: Remove the attention module: $\boldsymbol{\theta}_{base} \leftarrow \boldsymbol{\theta}_{base} - \boldsymbol{\theta}_{att}$
7: **for** each channel C with $I_C < \eta$ in model $\mathcal{M}_{base}$ do
8: Remove the channel C: $\boldsymbol{\theta}_{base} \leftarrow \boldsymbol{\theta}_{base} - \boldsymbol{\theta}_C$
9: **end for**
10: $\boldsymbol{\theta}_{pru} \leftarrow \boldsymbol{\theta}_{base}$, $\mathcal{M}_{pru} \leftarrow \mathcal{M}_{base}$
11: **return** Compressed model $\mathcal{M}_{pru}$ with parameters $\boldsymbol{\theta}_{pru}$

In summary, the "inner-model" optimization approach used in CroApp involves the use of attention modules to obtain importance scores, pruning the least important channels, and removing the inserted attention module to obtain the final pruned model. The attention module is designed with a spatial attention submodule and a channel attention submodule, arranged in a sequential "spatial-channel" fashion. The importance score vector I is calculated to decrease bias caused by input data, and channels are pruned based on a pre-defined pruning ratio.

5.3 Computation Sharing Across Multi-Application

Computation sharing aims to efficiently manage the shareable resources (e.g., CPU, GPU, memory) among all the concurrent applications [20, 24–27]. In the context of deep learning-enabled Industrial Internet, computation sharing allows the industrial equipment to share partial deep learning models for reducing resource consumption.

This approach can lead to significant benefits in terms of resource utilization, cost reduction, and improved performance. In this section, we will introduce the benefits of computation sharing and discuss how the CroApp framework leverages computation sharing to reduce resource consumption across various intelligent applications.

One of the most important benefits of computation sharing is the efficient utilization of computing resources. By sharing computing resources among multiple applications, enterprises can avoid the underutilization of resources that can occur when individual applications are assigned dedicated computing resources. This is especially important in industrial applications, where computing resources can be expensive and underutilized resources can result in significant cost overruns.

Another benefit of computation sharing is the ability to reduce costs. By sharing computing resources, organizations can reduce the number of resources needed to support their applications. This can result in significant cost savings, especially in situations where the organization has a large number of applications that require computing resources.

The third benefit of computation sharing is its potential to improve application performance. By sharing computing resources, organizations can ensure that applications have access to the computing resources they need to operate effectively. This can result in improved application performance, especially in situations where applications are competing for limited computing resources.

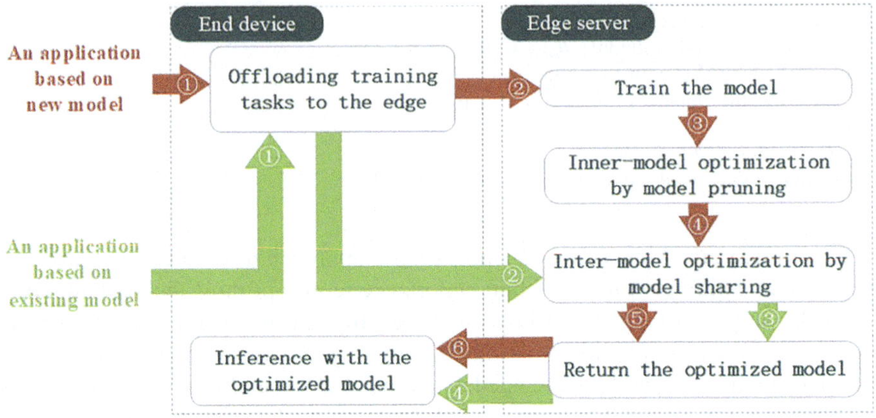

Fig. 5.3 The processing pipeline of CroApp when applications arrive

CroApp performs "inter-model" optimization on the edge server based on the results of "inner-model" optimization in a pipeline fashion. Figure 5.3 depicts the processing pipeline of CroApp when applications arrive. Where the "inner-model" optimization has been described in the previous section 5.2, and the "inner-model" optimization will be explained in the remainder of this section.

The goal of inter-model optimization is to automatically minimize redundant resource usage across applications. In this section, we will delve into the details of CroApp's "inter-model" optimization process.

First, CroApp trains a basic model by a common training process on the edge server for each kind of compressed model $\mathcal{M}_{\text{pru}}$. The trained model and its number of layers are denoted as $\mathcal{M}_{\text{tra}}$ and N, respectively. Once the model $\mathcal{M}_{\text{tra}}$ has been trained, CroApp segments it into two parts: a shareable model and a specific model, which can be formulated as follows:

$$\begin{cases} \mathcal{M}_{\text{tra}} & = m_f + m_u \\ N & = N_f + N_u \end{cases}, \tag{5.6}$$

where the shareable model m_f is consisted by the front several layers N_f and the specific model m_u is consisted by other layers N_u. The shareable model is a portion of the basic model that can be used by multiple compressed models, while the specific model is unique to each compressed model. This segmentation process is crucial for minimizing the storage requirements and computational complexity of the compressed models, enabling them to be deployed on resource-constrained edge devices.

Then CroApp generates a set of combinations of the shareable model and the specific model denoted as $\mathbb{S}$, where each combination has a different layer number of the shareable model and a specific model, which can be formulated as follows:

$$\begin{cases} \mathbb{S} & = \left\{ \mathcal{M}^1, \mathcal{M}^2, \ldots, \mathcal{M}^{\lceil \frac{N}{\sigma} \rceil} \right\}, \\ \mathcal{M}^t & = m_f^t + m_u^t, t \in \left\{ 1, 2, \ldots, \left\lceil \frac{N}{\sigma} \right\rceil \right\} \end{cases}, \tag{5.7}$$

where $\mathcal{M}^t$ represents a combination consisting by a shareable model with the front t groups of layers and a specific model with the other layers, σ is a coefficient to divide the N layers into $\lceil \frac{N}{\sigma} \rceil$ groups.

Next, CroApp fine-tunes each combination $\mathcal{M}^t$ for each application. Through this process, CroApp obtains the application's performance value $\text{per}(\mathcal{M}^t)$, as well as the storage resource consumption value $\text{sto}(\mathcal{M}^t)$ and computation resource consumption value $\text{com}(\mathcal{M}^t)$ of each combination. Afterward, CroApp selects the combinations $\mathbb{S}_{\text{opt}}$ that satisfy the Pareto-optimal frontier to ensure application performance and satisfy the resource budget.

Subsequently, CroApp adaptively chooses the combination with the maximum average performance of all applications in $\mathbb{S}_{\text{opt}}$ as the final result, which can be represented as follows:

$$\mathcal{M}^{\text{opt}} = \underset{\mathcal{M}^t}{\text{argmax}} \frac{1}{A} \sum_{q=1}^{A} \text{per}\left((\mathcal{M}^t)^{(q)} \right),$$
$$s.t. \begin{cases} \mathcal{M}^{\text{opt}} = m_f^{\text{opt}} + m_u^{\text{opt}} \\ B_s \geq \text{sto}\left(m_f^{\text{opt}} \right) + \sum_{q=1}^{A} sto\left((m_u^{\text{opt}})^{(q)} \right) \\ B_c \geq \sum_{q=1}^{A} \text{com}\left((\mathcal{M}^{\text{opt}})^{(q)} \right) \end{cases}, \tag{5.8}$$

where A represents the number of applications, B_s and B_c represent the storage and computation resource budget, respectively, and the superscript q represents the q-th application. In addition, the storage resource consumption of the shareable model m_f^{opt} only needs to be counted once due to it is stored once on the end device.

Finally, CroApp sends the final result $\mathcal{M}^{\text{opt}}$ to the end device for inference. Figure 5.4 depicts the approach of the inference process on the end device. To conduct inference for multiple applications (e.g., App1, App2, and App3 in Fig. 5.4) with the same CNN model structure, CroApp adopts a shared model for all three applications to obtain the intermediate results. Subsequently, CroApp adopts specific models to obtain the final results of each application, respectively.

In summary, CroApp searches each combination of models to obtain their performance, storage resource consumption, and computation resource consumption. It then selects the optimal combinations that meet the Pareto-optimal frontier and satisfy the resource budget. By adaptively choosing the combination with the maximum average performance, CroApp ensures optimal performance for all applications. In addition, CroApp adopts a shared model to obtain intermediate results for

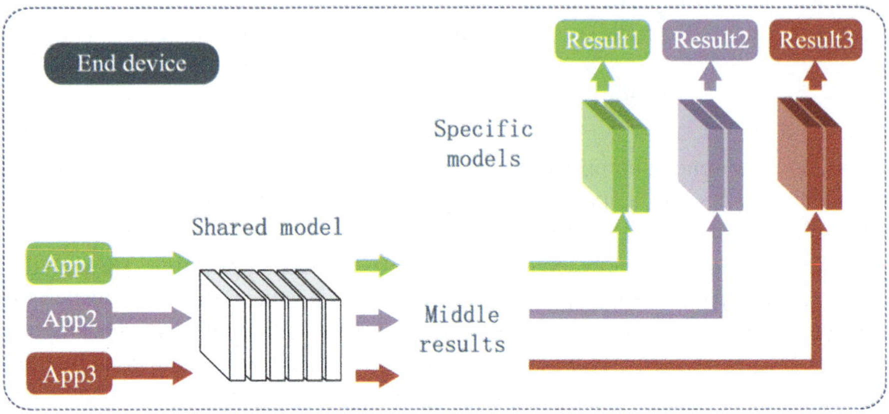

Fig. 5.4 Inference on the end device using a shared model and specific models

multiple applications with the same CNN model structure, reducing computational overhead and improving efficiency. Finally, CroApp uses specific models to obtain the final results for each application.

5.4 Deep Optimization for Industrial Applications

Deep optimization techniques have become increasingly popular in the industrial sector due to their ability to improve process efficiency, reduce costs, and enhance product quality. These techniques involve the application of deep learning algorithms to optimize complex industrial processes and systems. In this section, we will experimentally demonstrate the ability of CroApp framework for industrial applications in detail.

Experimental settings: We present a comparison of CroApp with state-of-the-art CNN-based resource optimization approaches in terms of both resource optimization performance and scalability. We utilize the widely adopted ResNet50 [28] model to simulate the model used by intelligent applications that are responsible for classification tasks. To evaluate the performance of CroApp, we use four popular public datasets, namely CIFAR10, CIFAR100, Fashion Mnist, and SVHN.

we use a GPU server equipped with an Intel Xeon Gold 5122 CPU and two RTX2080Ti GPUs as the edge server. In addition, we use a PC with an Intel Core i7-9750H CPU, NVIDIA Quadro T1000 GPU, and 16G RAM as the end device. These settings are commonly used in edge computing environments.

Ablation Experiments: In this section, we present the evaluation results of CroApp on "inner-model" optimization and "inter-model" optimization. As mentioned earlier, the performance of "inner-model" optimization is dependent on the pruning ratio, whereas the performance of "inter-model" optimization is influenced

Table 5.1 "Inner-model" optimization performance on CIFAR100 with different prune ratios. (The "ACC" represents "accuracy" and the "#para" in this table represents "number of parameters")

Strategy	Memory (#para)	Pruning ratio of #para	Computation (FLOPs)	Pruning ratio of FLOPS	ACC on CIFAR100	Improvement
Baseline	23.71 M	0%	2.60G	0%	75.88%	0%
CroApp-A	20.57 M	−13.2%	2.41G	−7.2%	76.43%	+0.37%
CroApp-B	17.70 M	−25.3%	2.23G	−14.2%	75.84%	−0.04%
CroApp-C	15.09 M	−36.3%	2.05G	−21.3%	75.43%	−0.45%
CroApp-D	12.68 M	−46.5%	1.86G	−28.3%	75.69%	−0.19%
CroApp-E	**10.66 M**	**−55.1%**	**1.84G**	**−29.2%**	**76.36%**	**+0.48%**
CroApp-F	8.54 M	−64.0%	1.50G	−42.2%	74.52%	−1.36%
CroApp-G	7.04 M	−70.3%	1.47G	−43.2%	74.73%	−1.15%
CroApp-H	4.64 M	−80.4%	1.18G	−54.6%	72.74%	−3.14%

by the number of frozen layers in the CNN model. We conducted experiments to evaluate CroApp's performance under different configurations.

To evaluate the performance of "inner-model" optimization, we ran an application that utilized ResNet50 on the end device. We used the Baseline to indicate the original resource consumption and accuracy of ResNet50. We set up 8 strategies with different pruning ratios and measured the parameters, FLOPs, and accuracy of the "inner-model" optimized model.

Table 5.1 presents the performance of "inner-model" optimization using CroApp. Our experimental results show that the CroApp is effective in reducing the unimportant parameters of the CNN model with no significant loss on accuracy. In addition, the accuracy of CroApp-A and CroApp-E is improved by 0.37% and 0.48%, respectively. This improvement is attributed to the fact that pruning can decrease the possibility of over-fitting. Interestingly, we observed that the accuracy is not linearly related to the pruning ratio, and the accuracy reached the highest point when we used the CroApp-E strategy. Therefore, we selected CroApp-E as our "inner-model" optimization strategy.

To evaluate the performance of "inter-model" optimization, we conducted experiments using 16 strategies. Each strategy corresponds to a combination of shared model and specific model discussed earlier. We used the CIFAR100 dataset to generate the shared model and CIFAR10, CIFAR100, Fashion Mnist, and SVHN datasets to generate specific models. We ran ten concurrent applications using ResNet50 on the end device to evaluate the average resource consumption and model accuracy.

Table 5.2 shows the performance of "inter-model" optimization using the CroApp-E strategy. Due to the "inter-model" optimization not being able to reduce FLOPs, we excluded it from Table 5.2. The "Shared blocks" represent the number of frozen blocks, where each block in ResNet50 has 3 layers. Our findings indicate that although the shared blocks are trained on CIFAR100, the accuracy of CroApp on CIFAR10, Fashion Mnist, and SVHN is still better than the original ResNet50 before we froze 7, 5, and 3 blocks, respectively. This finding shows that the features learned from CIFAR100 can be effectively transferred to other datasets. However, we observed that when the number of shared blocks exceeds 15, the accuracy on CIFAR10, Fashion Mnist, and SVHN decreases rapidly. This decrease in accuracy is due to most of the parameters in the model being frozen, which causes the model to lose its ability to fit the input data from these datasets. Furthermore, we found that the accuracy of CroApp on CIFAR100 had no significant loss and reached a higher point in most strategies. This result was due to the combination of the "inner-model" optimization and fine-tuning process.

In summary, our evaluation results show CroApp's capabilities in reducing resource consumption of industrial equipment over the Industrial Internet by removing the unimportant parameters of the CNN model with no significant loss in accuracy. Furthermore, we found that the performance of "inter-model" optimization is dependent on the number of frozen layers in the CNN model, and the features learned from one dataset can be effectively transferred to another dataset. These findings suggest that CroApp can be used to optimize the resource consumption and accuracy of CNN models across applications in edge computing environments.

Comparison Experiments: We present a comparative analysis of CroApp with several state-of-the-art CNN-based resource optimization approaches. We aim to evaluate the performance of CroApp and investigate its effectiveness in multi-application scenarios. Specifically, we compare CroApp with two "inner-model" optimization approaches, GN-based channel pruning approach (GNCP) [29] and CPSCA [15], as well as two "inter-model" optimization approaches, MCDNN [30] and Mainstream [24]. To simulate scenarios where different applications use the same model, we assume that there can be at most 10 applications using the same model. Whenever the number of applications exceeds 10, we assume that they use different models even if they actually use the same model.

In terms of resource optimization, the main consideration in multi-application scenarios is the trade-off between resource consumption and application performance. We consider a storage resource budget of 100 M (i.e., the total parameter number of the models on the device), which corresponds to the total parameter number of the models on the device. Figure 5.5 shows the average accuracy of each approach on the four datasets with varying numbers of applications. The baseline accuracy depicts the original ResNet50 model on the four datasets.

We observe that GNCP, CPSCA, and MCDNN exhaust the resource when the number of applications exceeds 7, 10, and 10, respectively, as they cannot adaptively adjust the resource consumption. In contrast, both Mainstream and CroApp can dynamically adjust the resource consumption during application execution. Although Mainstream does not use any model compression technique, its accuracy

Table 5.2 Inner-model optimization performance on CIFAR10, CIFAR100 fashion MNIST, and SVHN with different number of shared blocks. (The "ACC" in this table represents "accuracy" and the "#para" represents "number of parameters")

Strategy	Shared blocks	Average memory (#para)	Pruning ratio of #para	ACC on CIFAR10	ACC on CIFAR100	ACC on Fashion Mnist	ACC on SVHN	Average ACC
CroApp-E1	1	10.59 M	−55.34%	94.49%	76.4%	95.17%	96.1%	90.57%
CroApp-E2	2	10.53 M	−55.61%	94.71%	76.77%	95.13%	96.2%	90.68%
CroApp-E3	3	10.46 M	−55.88%	94.62%	76.54%	94.94%	96.26%	90.59%
CroApp-E4	4	10.26 M	−56.71%	94.51%	76.51%	94.63%	95.98%	90.41%
CroApp-E5	5	10.17 M	−57.09%	94.67%	76.72%	94.87%	95.99%	90.56%
CroApp-E6	6	10.08 M	−57.47%	94.48%	76.73%	94.58%	95.99%	90.45%
CroApp-E7	7	9.99 M	−57.89%	94.39%	76.4%	94.18%	95.77%	90.19%
CroApp-E8	8	9.20 M	−61.19%	93.83%	76.34%	93.19%	94.26%	89.41%
CroApp-E9	9	8.83 M	−62.75%	93.94%	75.7%	92.96%	94.14%	89.19%
CroApp-E10	10	8.46 M	−64.34%	93.73%	75.4%	93.1%	93.83%	89.02%
CroApp-E11	11	8.09 M	−65.90%	93.52%	75.69%	92.71%	93.85%	88.94%
CroApp-E12	12	7.72 M	−67.45%	93.02%	74.92%	92.62%	93.18%	88.44%
CroApp-E13	13	7.35 M	−69.01%	92.88%	75.91%	92.48%	91.63%	88.23%

(continued)

Table 5.2 (continued)

Strategy	Shared blocks	Average memory (#para)	Pruning ratio of #para	ACC on CIFAR10	ACC on CIFAR100	ACC on Fashion Mnist	ACC on SVHN	Average ACC
CroApp-E14	14	4.22 M	−82.22%	91.24%	75.97%	89.87%	73.84%	82.73%
CroApp-E15	15	2.73 M	−88.48%	89.68%	75.78%	88.72%	56.11%	77.57%
CroApp-E16	16	1.26 M	−94.71%	80.04%	76.15%	69.86%	25.28%	62.83%

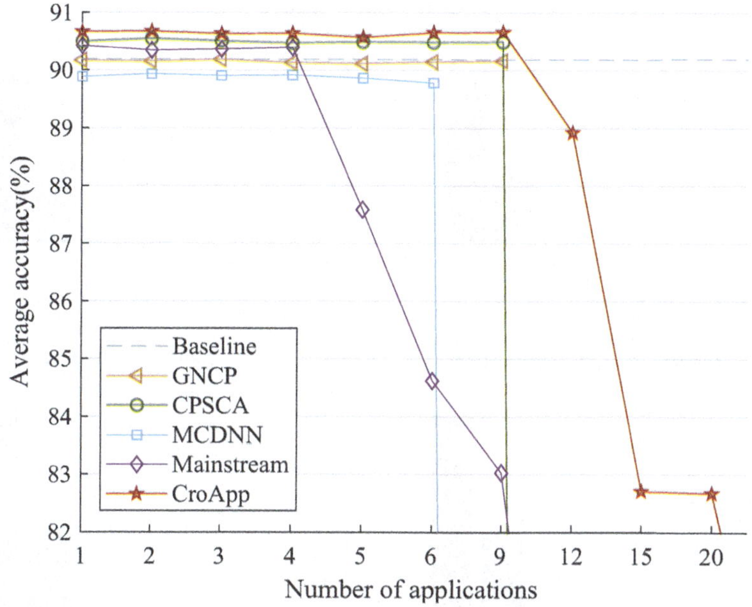

Fig. 5.5 Comparison of average accuracy with resource budget 100 M by GNCP, CPSCA, MCDNN, Mainstream, and CroApp

is lower than that of CroApp. Moreover, as the number of applications increases, CroApp maintains the best accuracy due to its "inner-model" optimization approach that reduces the risk of overfitting and its "inter-model" optimization approach that has a fine-tuning process to enhance the model's capability. In addition, CroApp adaptively sacrifices accuracy to trade-off resources as the number of applications increases.

To evaluate the scalability of different approaches, we vary the resource budget and application numbers. Figure 5.6 shows the maximum application number of each approach with varying resource budgets, whereas Fig. 5.7 shows the average accuracy of Mainstream and CroApp with varying application numbers.

GNCP and CPSCA have the same maximum application number because both adopt channel pruning for resource optimization, which results in the same resource consumption. MCDNN and Mainstream are model-sharing-based approaches, but Mainstream has better scalability because MCDNN cannot adaptively adjust resource consumption during execution. Compared to Mainstream, CroApp possesses both model compression and model sharing advantages, leading to better scalability and average accuracy.

In summary, the comparative analysis demonstrates that CroApp is an effective approach for resource optimization in multi-application scenarios, offering better scalability and accuracy than other state-of-the-art CNN-based resource optimization approaches.

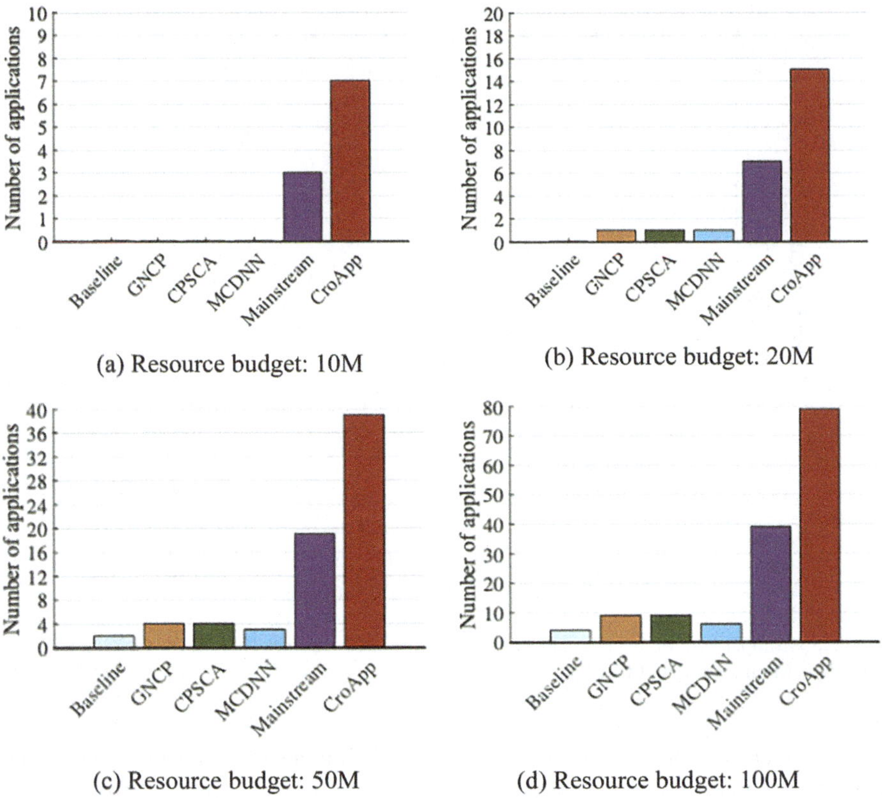

Fig. 5.6 Comparison of number of applications with different resource budgets by GNCP, CPSCA, MCDNN, Mainstream, and CroApp. (**a**) Resource budget: 10 M. (**b**) Resource budget: 20 M. (**c**) Resource budget: 50 M. (**d**) Resource budget: 100 M

5.5 Summary

Deep learning has emerged as a powerful tool for analyzing and processing vast amounts of data, and its applications are widespread across various domains, including the Industrial Internet. In this chapter, we have explored the several classical ways in which deep learning is leveraged to empower the Industrial Internet, along with the CroApp framework for saving storage and computational resources during training deep learning models.

We first discussed the benefits of deep learning in industrial applications, which include data-driven insights, improved efficiency, and optimization of operations. However, businesses need to overcome several challenges to successfully implement deep learning applications, including resource consumption, data quality and availability, and algorithm complexity. By investing in data quality and management

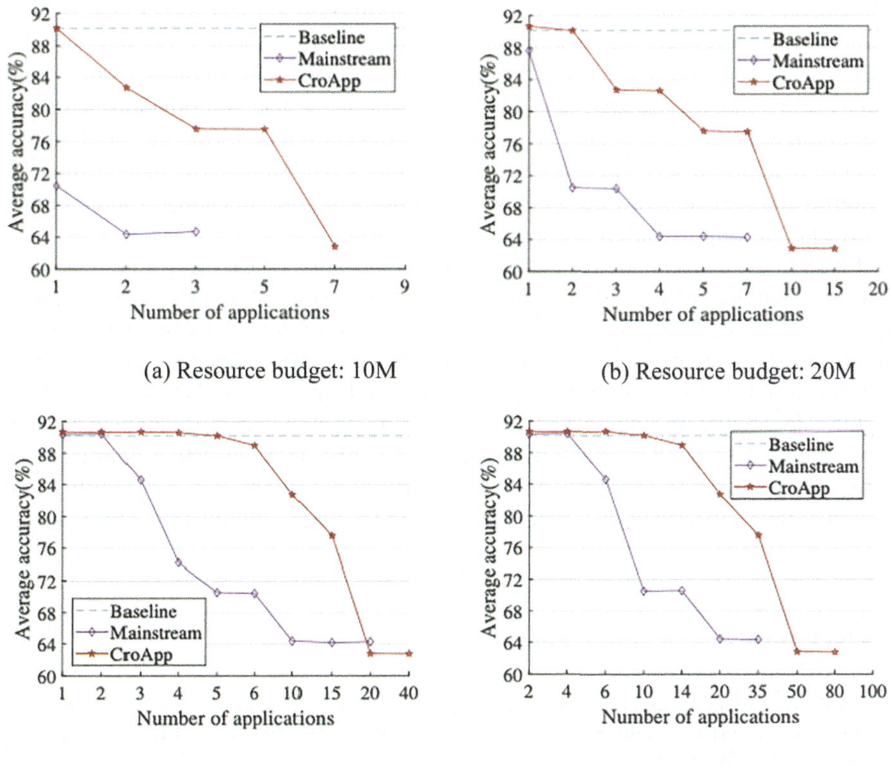

(a) Resource budget: 10M

(b) Resource budget: 20M

(c) Resource budget: 50M

(d) Resource budget: 100M

Fig. 5.7 Comparison of average accuracy with different resource budgets by Mainstream and CroApp. (**a**) Resource budget: 10 M. (**b**) Resource budget: 20 M. (**c**) Resource budget: 50 M. (**d**) Resource budget: 100 M

processes, cloud computing services, and training programs, businesses can effectively unlock the potential of deep learning in the Industrial Internet.

We then introduced attentive model compression for deep learning models, which is an effective way to reduce model size and computational complexity while maintaining high accuracy. We discussed the CroApp framework, which integrates attentive model compression for saving storage and computational resources caused by training deep learning models. The "inner-model" optimization used in CroApp involves the use of attention modules to obtain importance scores, pruning the least important channels, and removing the inserted attention module to obtain the final pruned model.

Next, we discussed computation sharing, which refers to the sharing of computing resources among multiple applications. This approach can lead to significant benefits in terms of resource utilization, cost reduction, and improved performance. We demonstrated how the CroApp framework leverages computation sharing to reduce resource consumption across various intelligent applications. Specifically, we described how CroApp searches each combination of models to obtain their performance, storage resource consumption, and computation resource consumption. It then selects the optimal combinations that meet the Pareto-optimal frontier and satisfy the resource budget. By adaptively choosing the combination with the maximum average performance, CroApp ensures optimal performance for all applications. In addition, CroApp adopts a shared model to obtain intermediate results for multiple applications with the same CNN model structure, reducing computational overhead and improving efficiency.

Finally, we experimentally demonstrated the ability of CroApp framework for industrial applications in detail. Our evaluation results showed that CroApp is effective in reducing the unimportant parameters of the CNN model with no significant loss in accuracy. We also found that the performance of "inter-model" optimization is dependent on the number of frozen layers in the CNN model, and the features learned from one dataset can be effectively transferred to another dataset. Our comparative analysis demonstrated that CroApp is an effective approach for resource optimization in multi-application scenarios, offering better scalability and accuracy than other state-of-the-art CNN-based resource optimization approaches.

References

1. Liang W, Yiyong H, Zhou X, Pan Y, Kevin I, Wang K (2021) Variational few-shot learning for microservice-oriented intrusion detection in distributed industrial IoT. IEEE Trans Industr Inform 18(8):5087–5095
2. Xu D, Li T, Li Y, Xiang S, Tarkoma S, Jiang T, Crowcroft J, Hui P (2021) Edge intelligence: empowering intelligence to the edge of network. Proc IEEE 109(11):1778–1837
3. Zhang C, Patras P, Haddadi H (2019) Deep learning in mobile and wireless networking: a survey. IEEE Commun Surveys Tutorials 21(3):2224–2287
4. Zhou X, Li Y, Liang W (2021) CNN-RNN based intelligent recommendation for online medical pre-diagnosis support. IEEE/ACM Trans Comput Biol Bioinform 18(3):912–921
5. Zhou X, Liang W, Shimizu S, Ma J, Jin Q (2021) Siamese neural network based few-shot learning for anomaly detection in industrial cyber-physical systems. IEEE Trans Industr Inform 17(8):5790–5798
6. Li Z, Yuan G, Niu W, Zhao P, Li Y, Cai Y, Shen X, et al. (2021) NPAS: a compiler-aware framework of Unified Network pruning and architecture search for beyond real-time mobile acceleration. In: Proceedings of the IEEE/CVF Conference on Computer Vision and Pattern Recognition, pp 14255–14266
7. Dupuis E, Novo D, O'Connor I, Bosio A (2020) On the automatic exploration of weight sharing for deep neural network compression. Des Automat Test Europe Conf Exhibit 2020:1319–1322
8. Shi W, Jie C, Quan Z, Li Y, Xu L (2016) Edge computing: vision and challenges. Internet Things J 3(5):637–646
9. Wang X, Han Y, Leung VC, Niyato D, Yan X, Chen X (2020) Convergence of edge computing and deep learning: a comprehensive survey. IEEE Commun Surveys Tutorials 22(2):869–904
10. Zhou X, Yang X, Ma J, Kevin I, Wang K (2021) Energy efficient smart routing based on link correlation mining for wireless edge computing in IoT. IEEE Internet Things J 9(16): 14988–14997

11. Cao K, Shiyan H, Shi Y, Colombo AW, Karnouskos S, Li X (2021) A survey on edge and edge-cloud computing assisted cyber-physical systems. IEEE Trans Industr Inform 17(11): 7806–7819
12. Tu X, Cheng X, Liu S, Li R, Xie G, Huang J, Yang LT (2021) Efficient monocular depth estimation for edge devices in internet of things. IEEE Trans Industr Inform 17(4):2821–2832
13. Fang W, Xue F, Ding Y, Xiong N, Leung VCM (2021) EdgeKE: an on-demand deep learning IoT system for cognitive big data on industrial edge devices. IEEE Trans Industr Inform 17(9): 6144–6152
14. Gong C, Ye L, Xie K, Jin Z, Li T, Wang Y (2021) Elastic significant bit quantization and acceleration for deep neural networks. IEEE Trans Parallel Distributed Syst 33(11):3178–3193
15. Liu M, Fang W, Ma X, Wenyuan X, Xiong N, Ding Y (2021) Channel pruning guided by spatial and channel attention for DNNs in intelligent edge computing. Appl Soft Comput 110: 107636
16. Kim K, Ji BM, Yoon D, Hwang S (2021) Self-knowledge distillation with progressive refinement of targets. In: Proceedings of the IEEE/CVF International Conference on Computer Vision, pp 6567–6576
17. Jia Y, Liu B, Dou W, Xu X, Zhou X, Qi L, Yan Z (2022) CroApp: a CNN-based resource optimization approach in edge computing environment. IEEE Trans Industr Inform 18(9): 6300–6307
18. Wang W, Chen M, Zhao S, Chen L, Hu J, Liu H, Cai D, He X, Liu W (2021) Accelerate Cnns from three dimensions: a comprehensive pruning framework. In: International Conference on Machine Learning, pp 10717–10726
19. Sui Y, Yin M, Xie Y, Phan H, Zonouz SA, Yuan B (2021) CHIP: channel independence-based pruning for compact neural networks. Adv Neural Inf Proc Syst 34:24604–24616
20. Guo J, Zhang W, Ouyang W, Dong X (2020) Model compression using progressive channel pruning. IEEE Trans Circuits Syst Video Technol 31(3):1114–1124
21. Jacob B, Kligys S, Chen B, Zhu M, Tang M, Howard A, Adam H, Kalenichenko D (2018) Quantization and training of neural networks for efficient integer-arithmetic-only inference. In: Proceedings of the IEEE/CVF Conference on Computer Vision and Pattern Recognition, pp 2704–2713
22. Chen Y, Li C, Gong L, Wen X, Zhang Y, Shi W (2020) A deep neural network compression algorithm based on knowledge transfer for edge devices. Comput Commun 163:186–194
23. Li X, Hu X, Yang J (2019) Spatial group-wise enhance: improving semantic feature learning in convolutional networks. arXiv:1905.09646
24. Jiang AH, Wong DL-K, Canel C, Tang L, Misra I, Kaminsky M, Kozuch MA, Pillai P, Andersen DG, Ganger GR (2018) Mainstream: dynamic stem-sharing for multi-tenant video processing. In: USENIX Annual Technical Conference, pp 29–42
25. Larsen JS, Clemmensen L (2020) Weight sharing and deep learning for spectral data. In: IEEE International Conference on Acoustics, Speech and Signal Processing, pp 4227–4231
26. Hung C-C, Ananthanarayanan G, Bodik P, Golubchik L, Minlan Y, Bahl P, Philipose M (2018) Videoedge: processing camera streams using hierarchical clusters. In: IEEE/ACM Symposium on Edge Computing, pp 115–131
27. Kim J-Y, Choi D-W (2021) Split-and-bridge: adaptable class incremental learning within a single neural network. Proc AAAI Conf Artif Intellig 35:8137–8145
28. He K, Zhang X, Ren S, Sun J (2016) Deep residual learning for image recognition. In: Proceedings of the IEEE/CVF Conference on Computer Vision and Pattern Recognition, pp 770–778
29. Wu Y, He K (2018) Group normalization. In: Proceedings of the European Conference on Computer Vision, pp 3–19
30. Han S, Shen H, Philipose M, Agarwal S, Wolman A, Krishnamurthy A (2016) Mcdnn: An Approximation-Based Execution Framework for Deep Stream Processing Under Resource Constraints. In: Proceedings of the 14th Annual International Conference on Mobile Systems, Applications, and Services, pp 123–136

Chapter 6
User Data Privacy Protection in Industrial Application

Abstract With the rapid development and widespread adoption of Internet of Things (IoT) technology, an increasing number of devices and objects are now interconnected, resulting in a substantial volume of data, especially the location data. Common components of a location-based service (LBS) include "location," "content," "timestamp," and "range." When it comes to safeguarding personal information, the location attribute and the content property are crucial. While most current privacy protection measures primarily focus on safeguarding the location attribute, they tend to overlook the content component of the LBS system. To tackle this challenge, we introduce a novel approach called content property privacy (CPP), which specifically leverages the content aspect to enhance privacy protection. To hide private information, the CPP approach makes use of k-anonymity to create fake content characteristics. An analysis of real-world data showed that the CPP technique may really improve privacy protection.

With the rapid development and widespread adoption of Internet of Things (IoT) technology, an increasing number of devices and objects are now interconnected, resulting in a substantial volume of data. This data encompasses various types of information, such as location data, environmental data, and sensor data. These data are widely used in smart home, intelligent transportation, industrial automation, health care, and other fields, which bring great convenience and benefits to people's life and work. Industrial Internet can realize the interconnection between equipment and factories, and improve the production efficiency and quality of factories through technical means such as data acquisition, analysis, and processing.

However, as data continues to grow and be utilized, concerns regarding personal privacy and data security have emerged.

Location information, especially in the context of location-based services (LBS), is of great concern due to its high sensitivity and accuracy. It is more likely to be abused or maliciously exploited. For example, attackers can use the obtained location information to track an individual's whereabouts, and collect sensitive information, resulting in the disclosure of personal privacy and security.

In addition to preserving personal privacy and data security, the Industrial Internet must acquire the location information of devices and objects to enable the

W. Dou et al., *Intelligent Industrial Internet Systems*,
https://doi.org/10.1007/978-981-99-5732-3_6

communication and interaction of devices. To guarantee the smooth operation of industrial applications and the preservation of user privacy, it is necessary to strike a balance between the requirements of service quality and user experience within the Industrial Internet.

Common components of a location-based service (LBS) include "location," "content," "timestamp," and "range." When it comes to safeguarding personal information, the location attribute and the content property are crucial. While most current privacy protection measures primarily focus on safeguarding the location attribute, they tend to overlook the content component of the LBS system. To tackle this challenge, we introduce a novel approach called content property privacy (CPP), which specifically leverages the content aspect to enhance privacy protection. To hide private information, the CPP approach makes use of *k*-anonymity to create fake content characteristics. An analysis of real-world data showed that the CPP technique may really improve privacy protection.

6.1 Location Data Privacy of User Over Industrial Internet

Location-based services (LBSs) are becoming essential to IoT applications. These services provide customers with customized suggestions and contextualized services depending on their location [1–4]. A user queries an LSP with their location and material of interest in an IoT-based LBS. The location service provider (LSP) then returns a point of interest (POI) near the user's location. Users can experience LBS convenience by engaging with the LSP.

Privacy problems may accompany these advantages brought by LBS. Untrusted LSPs [5] can employ LBS queries to deduce users' location, preferences, and health state. Therefore, users' privacy must be strengthened in IoT-based LBSs.

In recent years, location data privacy of user over Industrial Internet has attracted a lot of attention, and there have been numerous related research studies [2, 4, 6–15]. Among them, the *k*-anonymity-based dummy-based approach is popular [16]. This approach creates *k*–1 real-looking fake LBS queries for each user's query and mixes them with the user's real query to protect it. However, the existing dummy-based approaches only evaluate the loc property in the LBS query and neglect the content attribute, which might leak critical content information.

In this chapter, a content-aware privacy protection approach for IoT-based LBSs (CPP) is presented, which generate fake LBS queries that take *loc* and *content* aspects into consideration to protect privacy. The CPP technique uses hypothetical adversaries' background knowledge, such as query likelihood and reachability, to achieve theoretical *k*-anonymity.

This chapter shows how dummy-based approaches that ignore the content property might leak privacy. Existing approaches construct several dummy LBS requests using the *loc* property of the genuine LBS query and maintain the content attribute intact [16]. LBS query *loc* is connected to the content. Table tennis enthusiasts may be near a sports facility. In IoT-based LBSs, the CPP technique creates fake LBS queries that consider loc and content properties, improving user privacy.

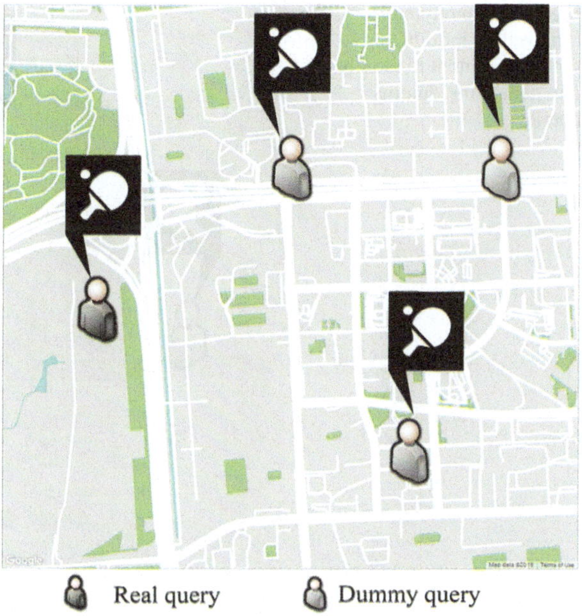

Fig. 6.1 4-anonymity achieved by introducing the location attribute into generating dummy queries

To achieve 4-anonymity, the current dummy-based technique creates three mock LBS inquiries depending on the loc attribute of the genuine LBS query (Fig. 6.1). Multiple content characteristics are identical. An opponent may value the content attribute over the loc attribute. The attacker can deduce the user's interests from their content attribute even when they don't know their location. Figure 6.1 shows that numerous LBS searches have various loc characteristics distributed over the map. All LBS searches have table tennis content characteristics.

Using the *k*-anonymity principle, users' content attributes may be concealed by selecting *k*–1 characteristics at random from the content database. The server then gets the fragmented data. This makes it less likely that an opponent will be able to ascertain the user's true content quality.

Figure 6.2 shows that randomly picking content qualities does not reduce likelihood. Table tennis, weightlifting, haircut, and yoga are the content characteristics of the four LBS searches (Fig. 6.2). The four qualities are semantically divided into sports and life services. 3/4 users need sports. In extreme circumstances, randomly picking *k*–1 content characteristics violates the *k*-anonymity principle. To meet the *k*-anonymity principle, while choosing *k*-1 content characteristics, examine their semantic importance.

New issues arise when considering the possibility of content queries for each site. Grids with different content query probabilities help grasp the map [6]. This section assumes that all grid content characteristics are the same for simplicity. Thus, grids have the same content query probability.

As illustrated in Fig. 6.3, the map is separated into grids with different colors indicating query probability. Borlange data generates these query probabilities. The

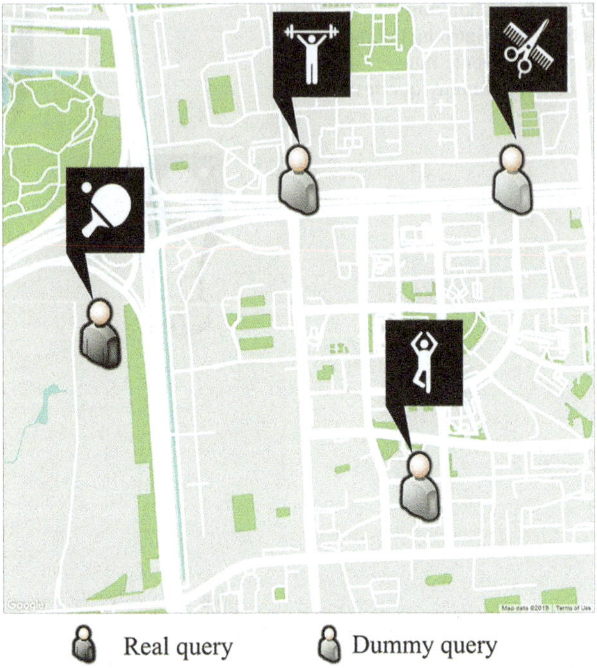

Fig. 6.2 4-anonymity achieved by introducing semantic content randomly into generating dummy queries

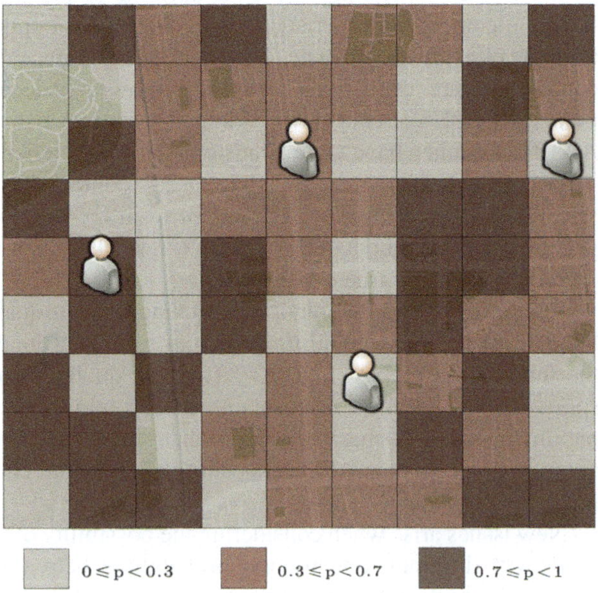

Fig. 6.3 Dummy queries generated by considering each grid's probability of initiating semantic content

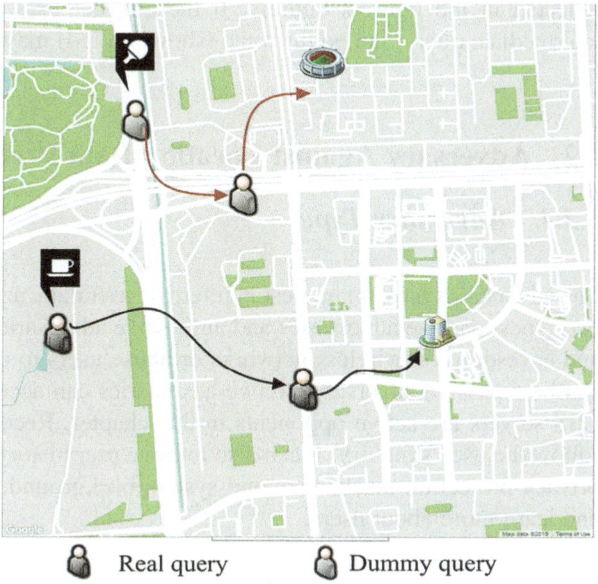

Fig. 6.4 Dummy trajectory piloted by introducing the user's mobility into generating dummy LBS queries

loc and content attributes fulfill the *k*-anonymity principle, and the existing approach may randomly produce *k*–1 dummy LBS searches for users. The attacker has the same relationship between the loc element and the content attribute, hence the likelihood of disclosing the genuine LBS query is 3/4. Specifically, the privacy protection impact is not 1/*k* when he knows the map grid's query probability. This chapter illustrates this issue with the following figure.

The figure's squares' colors show their content inquiry likelihood. In the figure, dummy LBS queries are created for the user independent of location and content connection, and their locations are dispersed everywhere. The opponent discovers that two of the four queries have a chance of 0.3 having LBS material. The other two have a probability of 0.3–0.7 and 0.7–1. The attacker can ignore the two requests with a probability of 0.3 and assume the user's LBS query is in the dark brown region. The user's actual LBS query exposure has fallen from 1/4 to 1/2.

Protectors confronting a single-query attacker will face the issues mentioned. The following paragraph considers the content–location relationship as the user moves.

Inquiry goals are often approached gradually by users [17]. In light of this, the suggested method would ignore the query target that is geographically closest to the user when the user conducts a continuous inquiry, supposing that the user will instead progressively approach the target. The user's real LBS query in Fig. 6.4 is to locate a coffee shop, while the sham inquiry requests information on table tennis facilities. Once the LSP query result is received, the user will choose a shopping center within a two-block radius. On the journey to the mall, the user may use LBS, and a question is asked halfway there. This means that a fake LBS inquiry taken halfway must be located far from the fitness centers. The gymnasium setting is not reflected in the figure's dummy. If the attacker uses continuity filtering on this intermediate query, the bogus route (shown in red) may be exposed.

6.2 Adversary Against Location Data Privacy of User

6.2.1 Adversary Type

The adversaries have an interest in inferring user data, and they can be classified into two types: passive adversaries and aggressive adversaries. If adversaries can follow and eavesdrop on wireless networks or abuse users to obtain sensitive user data, it can be a passive adversary. Active adversaries can access all data from the server. LBS servers are active opponents in this chapter. Recover global information and follow the user's inquiry. It can also retrieve user history. It also knows the location privacy protection mechanism and system background. It uses this data to deduce and learn more about users.

6.2.2 Adversary Behavior

First, we define the LBS query data owned by the adversary. The form of the data record is $(\mathrm{ID}, \mathrm{POI}_t)$, where ID refers to user feature identification ID, POI_t refers to user points of interest, and the composition form is $\{\{\mathrm{loc}_1, \mathrm{content}_1, t\}, \ldots, \{\mathrm{loc}_N, \mathrm{content}_N, t\}\}$; t is the user query time.

The attacker can utilize semantic recognition to identify dummy LBS queries after the user employs the existing dummy protection approach, decreasing the intended protection impact. The adversary also uses the continuity relationship of *content* in the dummy LBS to infer that there are some unreasonable continuous dummy LBS queries generated by the dummy method.

First, the semantic importance of location-based service (LBS) data is analyzed using a suitable methodology. As most queries are concise, LBS messages tend to be short. Typically, short texts contain around 100 words, such as Meituan search products. However, generating meaningful features from short texts poses a challenge in extracting linguistic features due to the limited sample size.

Rapid development updates LBS content. Traditional WordNet as the generic ontology may have coverage, information content update speed, and knowledge description issues. This chapter changes Wikipedia's ontology. Wikipedia has well-organized information. A directed graph connects each node on the page, showing outgoing and incoming relationships.

This chapter uses the Workload Management (WLM) technique to account for Wikipedia page link structure. First, the vocabulary is mapped to the equivalent Wikipedia term, and then the significance between words is derived by calculating the relevance of the mapped encyclopedia notion. A concept will have in-link and link-out links to other articles and concepts. The WLM algorithm analyzes these two

Step 1: Accept input: k(data points),K(number of clusters)
Step 2: construct similarity matrix $\in R^{k \times k}$
Step 2: construct matrix $P = D^{-1}WD^{\frac{1}{2}}$
Step 3: find the eigenvectors corresponding to the n largest eigenvalues of P,$v_1, v_2, \ldots, v_k$,
construct matrix $V = [v_1, v_2, \ldots, v_n] \in R^{n \times k}$
Step 4: normalize the row vector of V to get matrix Y, $y_{ij} = \frac{v_{ij}}{(\sum_j v_{ij}^2)^{\frac{1}{2}}}$
Step 5: regard each row of Y as a point, use Kmeans clustering
Step 6: output K clusters

Fig. 6.5 Steps of the clustering process

linkages and calculates the correlation between c_a and c_b. Instead of Google search results, the WLM algorithm uses Wikipedia links to calculate distance. The more ideas c_a and c_b share, the closer they are. Calculating semantic relevancy:

$$\mathrm{sr}(a,b) = \frac{\log\ \max(|A|,|B|) - \log|A \cap B|}{\log|W| - \log\ \min(|A|,|B|)} \tag{6.1}$$

The vector space model based on Term Frequency-Inverse Document Frequency (TF-IDF) is introduced by the WLM method to compute the link distance in Wikipedia. The following formula is used to determine how much each link should count: To determine the value of the connection from Article C to Article D, we use the following formula:

$$w(c \rightarrow d) = \log \frac{W}{D} \text{ if } c \in D \tag{6.2}$$

The aforementioned is the fundamental concept behind the WLM algorithm, which is quite developed and may be employed immediately. An adversary pre-trains a model that determines how important the phrases c_a and c_b are using the aforementioned WLM approach. The user may improve the adversary's ability to identify the LBS with only one query by specifying LBS = loc_1, $\mathrm{content}_1$, …, loc_k, $\mathrm{content}_k$, t. The correlation matrix-based clustering approach is applicable to the content information category.

Clustering function conceptual distance may be measured by correlation degree. Clustering distance is the correlation matrix. Figure 6.5 shows the clustering function.

The clustering function pre-builds a weighted undirected graph. Weight and graph division cluster text significance. Three stages comprise the function:

To determine the importance of various words, a weight matrix may be created using the WLM method. C_{ij} is an indicator of how important the words in sets i and j are.

D is a diagonal matrix built from C by adding up the items in the i-th column. Concurrently, make the matrix P equal to $D^{-1}WD^{\frac{1}{2}}$.

Assemble the n eigenvectors into the nk matrix V; It is recommended to do k-means clustering on each row of V separately.

This approach yields K clusters. The opponent sorts the K clusters by number of points from largest to smallest and picks the first ε clusters, believing they contain the user's true content information.

The opponent makes a grid out of the map cells and utilizes the information at hand to form a set. The set represents the likelihood that the cell will initiate a certain content inquiry. Here's how the composing format goes: cell $= \{\{\text{content}_1, p_1\}, \{\text{content}_1, p_1\}, \ldots, \{\text{Content}_n, p_n\}\}$, where $\{\text{content}_1, p_1\}$ means that the probability of launching content_1 in the current cell is p_1. The adversary integrates the content query probabilities of the entire map to form a matrix $\text{CELL} = \{\text{cell}_{ij}\}$, $i = 1 \ldots n, j = 1 \ldots n$.

The adversary uses CELL to filter out some LBS queries. Specifically, the filter is performed according to the following formula:

$$\begin{cases} \text{false}_{\text{LBS}} = \text{LBS} \mid \text{loc} \in \text{cell and } p_{\text{content}_{\text{loc}}} < \alpha \\ \text{true}_{\text{LBS}} = \text{loc} - \text{false}_{\text{LBS}} \end{cases} \tag{6.3}$$

The opponent, upon receiving the user's LBS query, first determines which cell the information pertains to, and then checks the cell's contents against the query probability. Some LBS queries are filtered out according to an adversary-customizable threshold and are considered dummy LBS queries if the corresponding content does not exist in the cell information and the semantic relevance is calculated, the query probability corresponding to the content information with the highest semantic relevance is selected.

The attacker can detect the fake LBS approach using query content continuity. Tu et al. hacked the user's GPS database and recovered 73–91% of user information, indicating a user movement trajectory. This suggests that the user will eventually return to his old trajectory. The protector sends the LSP k-1 mock LBS requests when the user makes a query at time t. The protection mechanism removes the LSP's genuine LBS query result.

6.3 CPP: A Content-Aware Privacy Protection Method

This chapter takes into account the continuity link between several LBS inquiries when the user performs a continuous inquiry and the relationship between loc information and content information in the LBS query while constructing a fake query to combat the aforementioned opponent.

To minimize semantic relevance of numerous query contents when users begin LBS searches, this chapter's solution must consider as much content information as

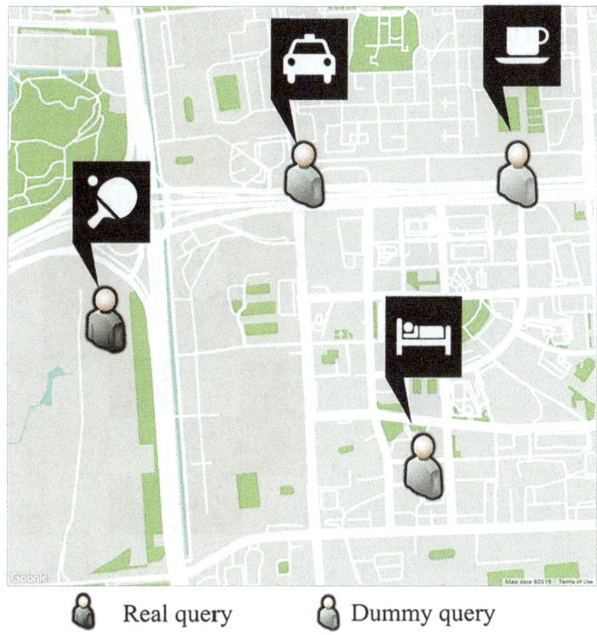

Fig. 6.6 Dummy queries generated by considering the otherness of the content attribute to achieve 4-anonymous

feasible while producing fake LBS questions. “Table tennis” is the genuine LBS query customers initiate, as seen in Fig. 6.6. Dummy LBS query content should be considered while constructing equivalent queries. The image shows cab, coffee, and hotel stuff.

When choosing the appropriate content based on the loc property, it is important to not only weigh the contents’ semantic similarity to one another but also to tailor our choice of content attribute to the loc attribute.

In Fig. 6.7, this chapter divides the map into an $n \times n$ grid, where each grid contains the query probabilities of multiple content. The composition of the figure is as follows.

Using a $n \times n$ grid, this section of the book displays the query probabilities of various material in Fig. 6.7. Here’s how the diagram is put together:

$$\text{Cell} = \{\{\text{content}_1, p_1\}, \{\text{content}_2, p_2\}, \ldots, \{\text{content}_n, p_n\}\} \tag{6.4}$$

where content is the set of words that make up the query (e.g., “coffee” or “parking lot”) and p is the likelihood that this query will be initiated in this grid. The matrix provides an expression for the likelihood of content queries throughout the whole map, as follows:

Fig. 6.7 4-anonymity achieved by introducing the content attribute possibility into generating dummy queries

$$\text{Grid} = \begin{bmatrix} \text{cell}_{11} & \cdots & \text{cell}_{1n} \\ \vdots & \vdots & \vdots \\ \text{cell}_{n1} & \cdots & \text{cell}_{nn} \end{bmatrix} \tag{6.5}$$

To improve the clarity of the expression in Fig. 6.7, this chapter combines several of the cell's content attributes into one. The probability that the grid contains the content query is expressed as follows: The probabilities of the content query in the grid are represented by the various colors of the grid when cell = cell = $\text{content}_1, p_1$ is used. As depicted in Fig. 6.7, the probability p corresponding to the launching of content in a user's actual LBS query ranges between 0.70 and 1. Following the generation of the loc content in the synthetic LBS query, the content with the content query probability in the corresponding cell and the content query probability in the actual LBS are selected in the same interval. Then, the semantic relevance test shown in Fig. 6.6 is executed to generate the corresponding dummy LBS query.

The preceding study considers LBS loc attribute and content attribute continuity in a single user query. In actual life, users usually start many searches. In a continuous inquiry, this chapter must address the content–loc connection.

In this chapter, we make the assumption that the user will gradually approach the location service provider (LSP) and receive the query result related to A after initiating a query specifically for target A. This process is facilitated by utilizing user orientation. For instance, the user may request data on café content content_a at a certain location loc_a. The LSP will provide information about local cafes after receiving inquiry A from the user, and the user may then choose their chosen café. The user demonstrates the trend of the LBS query in this procedure.

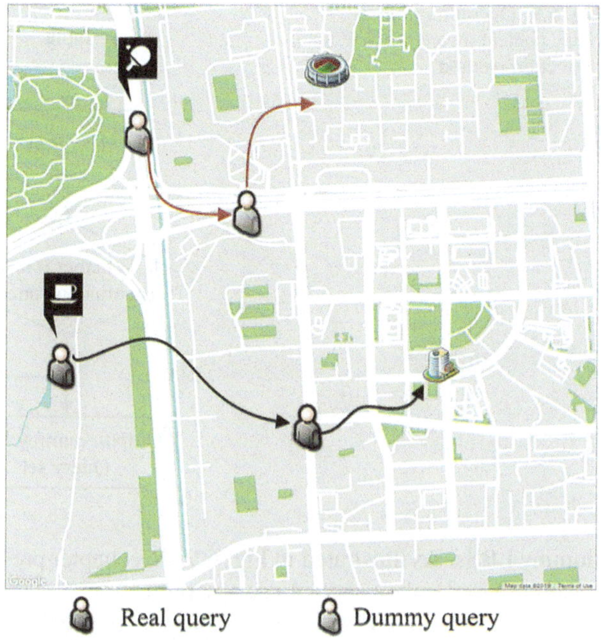

Fig. 6.8 Dummy trajectory piloted by introducing the reachability of the content attribute into generating dummy queries

In Fig. 6.8, the user's first actual LBS inquiry is about a coffee establishment. We generate a table tennis-themed mock LBS query for the user using the approach in Figs. 6.6 and 6.7. To simplify the problem, this chapter argues that the user can choose the highest-scoring cafe directly, and as the user moves closer to the cafe, the user will initiate a continuous LBS query. The LSP returns two queries for users to select the corresponding results of the real LBS check (i.e., near the cafe). This chapter's algorithm for generating dummy LBS queries is designed to converge on the query result (gym) of the preceding dummy LBS location, with a bias angle that is as near as feasible to that of a genuine continuous LBS query. The idea for making a phony LBS that is always on is shown in Fig. 6.8. During the course of the process, a false LBS forms and gradually moves toward the stadium, with a horizontal angle that mirrors that of the uninterrupted real LBS. This method safeguards users' repeated LBS requests.

In summary, this chapter considers the relationship between the content property and the loc attribute in the LBS query and the continuation of the inquiry when the user travels continually to build the most realistic fake LBS question for the user.

Cellular operators hold user-owned LBS query data and provide privacy protection services. LBS = loc, content composes the LBS query. This chapter's dummy LBS must examine loc and content properties.

This chapter introduces two algorithms for location-based content-aware privacy protection: Algorithms 6.1 and 6.2. This approach generates the task input according to the user's privacy needs, selectively performs Algorithms 6.1 and 6.2 depending on whether the user's inquiry is continuous, and ultimately obtains a collection of

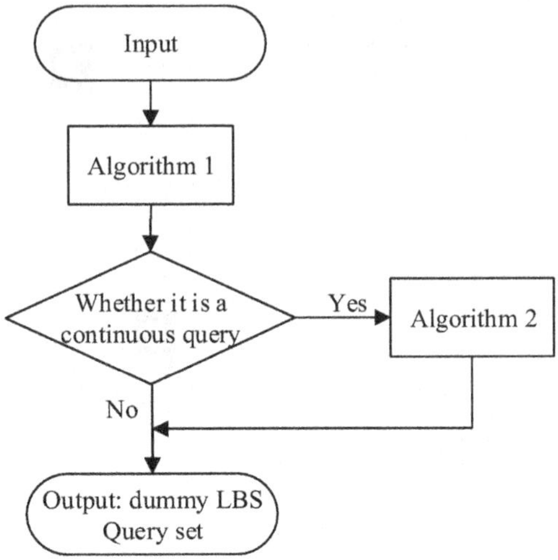

Fig. 6.9 Operation process of the content-aware privacy protection method

dummy LBSs. As illustrated in Fig. 6.9, this chapter presents a running flowchart for location-based online service privacy protection:

Algorithms 6.1 and 6.2 are as follows. In Algorithm 6.1, $\mathrm{Loc_{temp}}$ is generated using the content background knowledge of the cell where loc is located. After the user initiates a real $\mathrm{LBS_{true}} = \mathrm{loc_{true}}, \mathrm{content_{true}}$ query, first the algorithm of Liu is used to generate $\mathrm{Loc_{cand}}$, $\mathrm{Loc_{cand}} = \mathrm{loc_1}, \mathrm{loc_2}, \ldots, \mathrm{loc}_n$, where n is generally $2k$, and k is the user's privacy requirement. Each loc is then assigned to a cell based on its physical location after further curation. Here's what each cell's worth of data looks like: $\mathrm{cell} = \{\{\mathrm{content_1}, p_1\}, \{\mathrm{content_2}, p_2\}, \ldots, \{\mathrm{content}_n, p_n\}\}$. Calculate the semantic relevance $\mathrm{sr(content, content_{true})}$ between each content in the cell and $\mathrm{content_{true}}$, and then select the largest sr value

Algorithm 6.1: Generate a Dummy Query Considering the Relationship Between Location and Content

Input: LBS_{true}(the user's actual query), *Grid*(content probabilities for the cell of entire map), n(the number of loc_{temp})
Output: LBS_{temp}
1: generate the initial n candidate LBS LBS_{temp} by using Algorithm 6.2
2: $LBS_{dummy} = \phi$
3: for each $loc_{t+1} \in loc_{cand}$ do
4: calculate the *cell* for *loc* according the Grid Map rule
5: for each $content \in cell$ do
6: calculate $sr(content, content_{true})$, select the smallest *content*
7: add $(loc, content)$ in LBS_{temp}
8: end for
9: end for
10: return LBS_{temp}

Algorithm 6.2: Generate a Dummy Query Considering the time Reachability

Input: LBS_t(the LBS query initiated by the user at time t), α(customizable similarity requirements), β(customizable angle requirements), k(user privacy requirements)
Output: LBS_{dummy}
1: construct directed graphs G_{loc}, $G_{content}$
2: $LBS_{dummy} = \phi$
3: for each $loc_{t+1} \in G_{loc}$ do
4: check loc_{t+1} according to formula 6.6
5: end for
6: if $|loc_{t+1} \geq 2k|$ then
7: for each $content_{t+1} \in G_{content}$ do
8: check loc_{t+1} according to formula 6.7
9: add loc, $content$ into LBS_{dummy}
10: end for
11: else
12: run line 1-13
13: end if
14: return LBS_{dummy}

The process of generating a dummy LBS query at $t + 1$ is as follows. First, $\text{LBS}_{\text{temp}_{t+1}}$ is generated based on Algorithm 6.1, and the loc and content in LBS_t and $\text{LBS}_{\text{temp}_{t+1}}$, respectively, are transformed into a directed graph $G_{\text{loc}} = \{V_{\text{loc}}, E_{\text{loc}}\}$, $G_{\text{cont}} = V_{\text{cont}}, E_{\text{cont}}$; for loc $\in G_{\text{loc}}$, the following requirements should be met:

$$\begin{cases} |\text{loc}_{t+1}| \geq k \\ \forall \text{loc}_{t+1}, \exists \text{loc}_t : \angle < \text{loc}_{t+1}, \text{loc}_t > \geq \beta \end{cases} \tag{6.6}$$

$\angle < \text{loc}_{t+1}, \text{loc}_t>$ is the horizontal angle between $<\text{loc}_{t+1}, \text{loc}_t>$.

After filtering out the loc_{t+1} that meets the requirements, according to loc_{t+1} corresponding to content_{t+1}, content_{t+1} in G_{cont} should meet the following requirements:

$$\begin{cases} |\text{content}_{t+1}| \geq k \\ \forall \text{content}_{t+1}, \exists \text{content}_t : \text{sr}(\text{content}_{t+1}, \text{content}_t) \geq \alpha \end{cases} \tag{6.7}$$

$\text{sr}(\text{content}_{t+1}, \text{content}_t)$ is the semantic correlation between $(\text{content}_t, \text{content}_{t+1})$.

The computational cost of the content privacy protection approach is only partially taken up by the semantics computing of the fake LBS. Pre-training of the semantic relevance calculation model allows for its subsequent installation on the user's device. The semantics relevance check begins with a content reasonability judgment that narrows down a pool of n potential candidate dummies. In this step, at most $k_{i-1} \cdot n + k_{i-1} \cdot (n + 1)$ times computations are performed, and its upper bound of computation complexity is $O(k_{i-1} \cdot n)$. If all the initial candidate dummies pass the content reasonability judgment and form $C^1_{k_{i-1}} \times C^1_{n+1} = k_{i-1} \cdot (n+1)$ movement paths, then the comparison of time reachability requires computation complexity $k_{i-1} \cdot (n+1)$ times. As a result its upper bound of computation complexity is $O(k_{i-1} \cdot n)$, and its upper bound of computation complexity is $O(k_{i-1} \cdot n) + O(k_{i-1} \cdot n) = O(k_{i-1} \cdot n)$.

6.4 Analysis of CPP Applied to Industrial Application

6.4.1 Experiment Setup

An Intel Core i7-8550U processor (8 CPUs, 1.80 GHz) and 16 GB RAM host all tests on Windows.

6.4.1.1 Experimental Data

This section uses Microsoft Research's GeoLife trajectory data [18]. GeoLife collected GPS trajectory data for 178 users from April 2003 to October 2011. 17,621 trajectories cover 1,251,654 km and 48,203 h in this data collection. GPS trajectories are time-stamped. Each point has latitude, longitude, and height. These trajectories use GPS and GPS phone recordings with various sample rates. This data set records the user's outside sports, including shopping, sightseeing, dining, hiking, cycling, and lifestyle routines like coming home to work.

6.4.1.2 Experimental Metrics

Uncertainty λ

After privacy protection, entropy represents the likelihood of detecting actual LBS inquiries from all queries. This section divides the map into $N \times N$ cells with a query probability (q) of having been searched in the past: $\sum_{i=1}^{N^2} q_i = 1$ denotes the probability that an LBS query is $p_i = -\frac{q_i}{\sum_{j=1}^{m} q_i}$. The following expression defines the entropy H at which the genuine LBS is recognized in the anonymous set:

$$H = -\sum_{i=1}^{k} p_i \log_2 p_i \tag{6.8}$$

Since the entropy H is inversely proportional to the privacy effect, the purpose of this chapter is to ensure the greatest possible entropy. When k queries have identical p_i, the entropy is greatest and the value is $H_{\max} = \log_2^{k^2}$.

However, the above entropy does not take into account that after the online adversary receives k LBS queries processed by the privacy protection method, the adversary filters the k LBSs, and filters out m LBS queries Q_m, leaving n queries Q_n, referring to the uncertainty proposed by Niu et al., where λ is defined as follows:

$$\lambda = \begin{cases} \sum_{q \in Q_m} H_q & \text{if true}_{\text{loc}} \in Q_n \\ \log_2 k^2 & \text{otherwise} \end{cases} \# \tag{6.9}$$

According to formula (6.9), in order to increase the uncertainty λ, it is necessary to minimize the number of queries rejected by the adversary or to increase the entropy of the remaining queries Q_n.

The Proportion of Recognized Dummy LBS $\text{Rec}_{\text{Dummy}}$

$\text{Rec}_{\text{Dummy}}$ is a metric that indicates how many phony LBS inquiries the user's n queries contained. The procedure of calculation is as follows:

$$\text{Re}\, c_{\text{Dummy}} = \frac{1}{n} \sum_{i=0}^{n} \frac{|\text{false}_{\text{lbs}} \in \text{dummy}_{\text{lbs}}|}{\text{dummy}_{\text{lbs}}} \tag{6.10}$$

$\text{dummy}_{\text{lbs}}$ refers to the dummy LBS in the LBS inquiry after the privacy protection procedure has been applied, and $\text{false}_{\text{lbs}}$ refers to the query that the adversary considers to be a fictitious LBS.

The Proportion of Restored Trajectory Res_{Tra}

Res_{Tra} is a measure of the fidelity of trajectory restoration. Referring to the concept of Quattrone, this chapter encodes the points of the top k candidate paths in GeoHash and divides the trajectory into grid borders using GeoHash technology.

$$\text{Re}\, s_{\text{Tra}} = \frac{1}{n} \sum_{i=1}^{n} \sum_{j=1}^{k} \frac{1}{k} \frac{|\text{cand}_{\text{path}_i} \cap \text{daily}_{\text{path}_i}|}{|\text{daily}_{\text{path}_i}|} \tag{6.11}$$

When calculating Res_{Tra} in this chapter, calculate the number of grid repetitions between the k highest ranked candidate paths $\text{cand}_{\text{path}_i}$ and the j–th $\text{daily}_{\text{path}_j}$, divided by the number of grids occupied by the $\text{cand}_{\text{path}_i}$ trajectory.

Calculate the number of grid repetitions between the k highest ranked candidate paths $\text{cand}_{\text{path}_i}$ and the j–th $\text{daily}_{\text{path}_j}$, divided by the number of grids occupied by the $\text{cand}_{\text{path}_i}$ trajectory, when calculating Res_{Tra} in this chapter.

Comparison Methods

This section compares three dummy protection approaches with the content-aware privacy protection method to assess efficacy:

1. Baseline: This procedure is a method of random selection. When a dummy LBS query is generated for the user, the LBS is selected at random so that H receives the utmost value $H_{\max}$.
2. SCADBM [19]: First, the current dummy-based technique generates the first candidate dummy elements. Then, the authors examine the temporal and geographical connection between the provided ongoing LBS sets using time reachability, direction similarity, and in/out degree.
3. LODS [8]: To achieve these three aims, carefully choose k potential items from the local map: privacy, QoS, and long-term surveillance attacks.

Experimental Parameters

This experiment varies two privacy protection effect-related characteristics.:

1. Privacy Requirement: The value k reflects the privacy needs of consumers. The greater the user's desired level of privacy protection, the greater the value of k. The experimental value of k is $k = 4,6,8,10,12,14,16,18,20,22,24,26$.
2. The spectrum of acceptable semantic relevance sr: sr reflects the relevance of the content attribute in the synthetic and actual LBS queries. The greater the range sr, the greater the correlation between multiple content attributes, or the greater the adversary's ability to identify the target. The experimental value of sr is sr = 0.1,0.2,0.3...,1.

Based on the experimental configuration described above, this section undertakes three sets of experiments using the method proposed in this chapter and the three comparison methods of SCADBM, LODS, and Random. The three experimental groups all utilized the control variable method to determine the effect of the privacy requirement k and the allowable semantic relevance range sr on the uncertainty λ, the ratio of the identified dummy LBS $\mathrm{Rec}_{\mathrm{Dummy}}$, and the ratio of the recovered trajectory $\mathrm{Res}_{\mathrm{Tra}}$. The specific parameter configurations are shown in the following table (Table 6.1).

Table 6.1 Setting of experimental parameters

Experimental parameter	Privacy requirement k	The allowable range of semantic relevance sr
Privacy requirement k	4, 6, 8, . . ., 26	0.2
The allowable range of semantic relevance sr	12	0.1,0.2,0.3, . . ., 1

6.4.2 Comparison Results

Users can set the privacy requirement *k* and semantic relevance range sr based on their situation in real life. Figures 6.11 and 6.12 illustrate the experimental outcomes.

The uncertainty-based first set of experiments contrasts the four strategies. Figure 6.10 displays the uncertainty magnitude. Figure 6.10a illustrates the uncertainty magnitude as *k* grows. The chart shows that when anonymous demand rises, Baseline, SCADBM, LODS, and CPP techniques generate fake queries and uncertainty rises. Figure 6.10a shows that this chapter's CPP technique has maximum uncertainty. Figure 6.10b displays uncertainty magnitude as semantic relevance sr grows. The figure shows that when the semantic relevance range rises, the Baseline method, SCADBM uncertainty, and LODS uncertainty remain unchanged. This

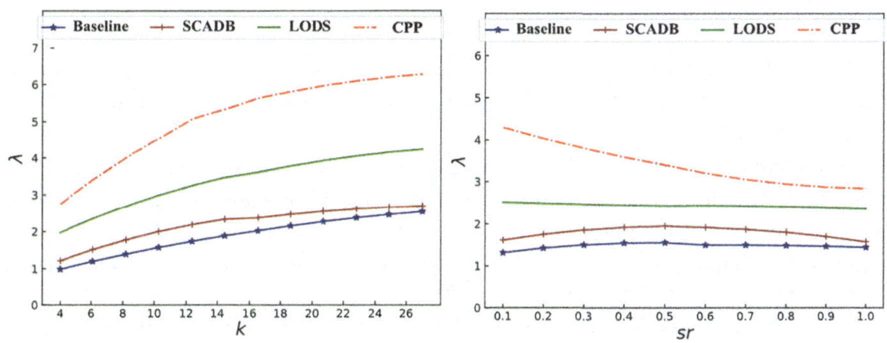

(a) (b)

Fig. 6.10 Impact of the privacy requirements *k* and semantic relevance range sr on λ. (**a**) *k* vs. λ. (**b**) sr vs. λ

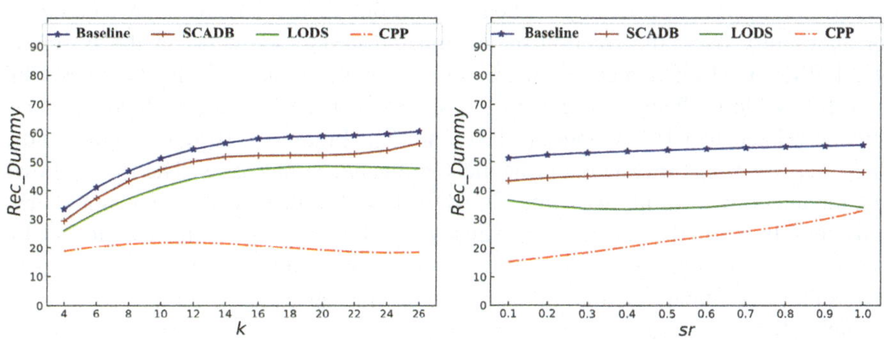

(a) (b)

Fig. 6.11 Impact of the privacy requirements *k* and semantic relevance range sr on Rec_{Dummy}. (**a**) *k* vs. Rec_{Dummy}. (**b**) sr vs. Rec_{Dummy}

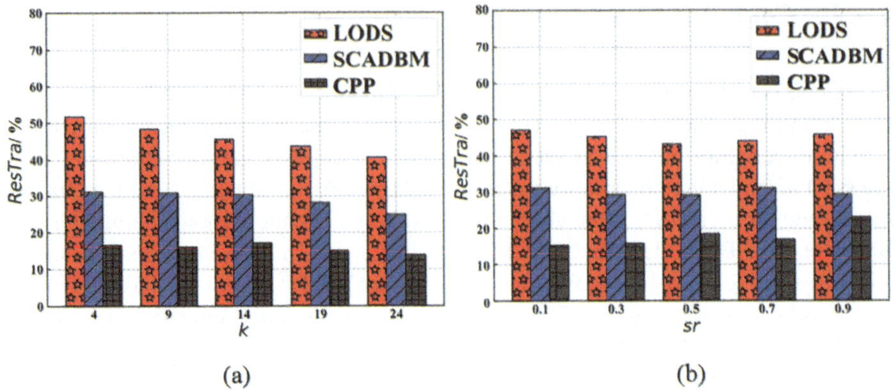

Fig. 6.12 Impact of the privacy requirements k and semantic relevance range sr on Res_{Tra}. (**a**) k vs. Res_{Tra} . (**b**) sr vs. Rec_{Tra}

chapter's CPP method's uncertainty is lowering but stays ideal. The procedure in this chapter has always maintained optimal performance, or the uncertainty of the greatest value of.

This chapter's CPP method generates mock LBS queries using the content property. The opponent is less likely to notice the CPP approach as k increases. This chapter's CPP technique optimizes uncertainty performance. The CPP technique in this chapter maintains optimal performance as the semantic relevance range rises, but sr reduces it. The content property cannot fulfill k-anonymity criteria since the CPP approach has steadily reduced the difference between content attributes in many fake LBS queries as sr increases. Consider this before setting sr. This chapter sets sr to 0.2. The other three methods ignore the semantic significance of the content property, therefore they are insensitive to the argument sr.

The second set of trials examined the four approaches' $\text{Rec}_{\text{Dummy}}$ LBS ratio performance. Figure 6.11 illustrates $\text{Rec}_{\text{Dummy}}$'s size. Figure 6.11a demonstrates how $\text{Rec}_{\text{Dummy}}$ fraction changes when anonymous demand k grows. As the desire for anonymity rises, the fraction of recognized dummy elements in the Baseline, SCADBM, and LODS methods increases and finally flattens. This chapter's CPP approach seldom changes. Figure 6.11a shows that this method's $\text{Rec}_{\text{Dummy}}$ percentage related to CPP is optimal, i.e., minimal. Figure 6.11b demonstrates how $\text{Rec}_{\text{Dummy}}$ fraction varies when semantic significance sr increases. The figure shows that the uncertainty of the Baseline, SCADBM, and LODS methods has not altered considerably while the semantic relevance range has increased. This chapter's CPP method dummy components are growing. It is comparable to LODS. Figure 6.11 shows that this chapter's strategy always yields the lowest proportion of recognized dummy LBS $\text{Rec}_{\text{Dummy}}$.

This chapter's CPP method generates a dummy LBS query that fully considers the content property. The adversary is less likely to recognize the CPP than the other three approaches as the anonymity requirement k increases. This chapter's CPP approach remains stable while other ways fail. As the permissible range of semantic significance expands, the CPP technique in this chapter performs optimally but progressively decreases until it virtually matches the LODS method. It is like the first group. The CPP technique gradually minimizes the difference between content attributes in repeated fake LBS queries until the content attribute cannot fulfill *k*-anonymity. Thus, this must be addressed while choosing the parameter sr size.

The third set of experiments compares the performance of the three methods from the perspective of the restored trajectory ratio Res_{Tra}. The ordinate in Fig. 6.12 represents the size of the restored trajectory ratio Res_{Tra}. Fig. 6.12a shows the restored trajectory ratio Res_{Tra} as the anonymous demand *k* increases. It can be observed from the figure that as the demand for anonymity continues to increase, the trajectory ratio Res_{Tra} restored by the Baseline method, SCADBM method, and LODS method is gradually decreasing, but it is still higher than the CPP method in this chapter. The CPP method in this chapter is relatively stable and almost unchanged. It can be observed from Fig. 6.12a that the proportion of restored trajectories Res_{Tra} corresponding to CPP in this chapter has been kept optimal, that is, the smallest. Figure 6.12b shows how the restored trajectory ratio Res_{Tra} changes as the allowable range of semantic relevance *sr* increases. It can be observed from the figure that as the allowable range of semantic relevance continues to increase, the proportions of restored trajectories of the Baseline method, SCADBM method, and LODS method have not changed significantly. The proportion of identified dummy elements in the method CPP of this chapter is gradually increasing, but it still has the smallest value. It can be observed from Fig. 6.12 that the method in this chapter has always maintained the optimal performance, namely, the proportion of restored trajectories Res_{Tra} is the smallest.

The third set of experiments examines the efficacy of the three methodologies based on the ratio of restored trajectory Res_{Tra}. Fig. 6.12's ordinate represents the magnitude of the restored trajectory ratio Res_{Tra}. As the anonymous demand *k* increases, Fig. 6.12a depicts the restored trajectory ratio Res_{Tra}. As shown in the graph, as the demand for anonymity increases, the trajectory ratio Res_{Tra} restored by the Baseline method, SCADBM method, and LODS method decreases progressively, but is still greater than the CPP method in this chapter. In this chapter, the CPP method is relatively stable and nearly unaltered. Figure 6.12a demonstrates that the proportion of restored trajectories Res_{Tra} corresponding to CPP in this chapter has been kept optimal, i.e., as small as possible. Figure 6.12b illustrates how the restored trajectory ratio Res_{Tra} varies as the acceptable range of semantic relevance sr expands. As the allowable range of semantic relevance increases, the proportions of restored trajectories for the Baseline method, SCADBM method, and LODS method have not changed significantly, as shown in the graph. The proportion of identified dummy elements in this chapter's CPP method is progressively increasing, but it still has the lowest value. Figure 6.12 demonstrates that the method presented in this chapter has always maintained the optimal performance, as measured by the minimum proportion of restored trajectories Res_{Tra}.

The final set of trials compares the three approaches using the restored trajectory ratio Res_{Tra}. In Fig. 6.12, Res_{Tra} is the ordinate. Figure 6.12a displays Res_{Tra} as *k* grows. The figure shows that when anonymity need rises, the trajectory ratio Res_{Tra}

restored by the Baseline technique, SCADBM method, and LODS method decreases, but it is still greater than the CPP approach in this chapter. This chapter's CPP approach remains stable. Figure 6.12a shows that the fraction of restored trajectories Res_{Tra} corresponding to CPP in this chapter has been maintained ideal, that is, the least. Figure 6.12b demonstrates how Res_{Tra} evolves as sr grows. The figure shows that when the permissible range of semantic significance increases, the proportions of restored trajectories of the Baseline, SCADBM, and LODS methods have not altered much. This chapter's method CPP has the fewest recognized dummy components, but it is growing. Figure 6.12 shows that this chapter's strategy always yields the lowest fraction of restored trajectories Res_{Tra}.

6.5 Summary

LBS query privacy has been prioritized in recent years. These attempts often use dummy-based methods. These techniques fail because they don't aware the opponent has side information and produce ignorance in a lake or hard mountain environment, which the opponents see as ineptitude. Since opponents will employ side material, researchers deliberately chooses dumb sites. The technique was improved by choosing far-apart dumb sites based on the entropy measure. Others collected more comments and minimized privacy leakage for each individual questionnaire position and the dumb position. To decrease terminal database server requests, stored data was used to answer questions. Terminal security limits terminal query server inquiries.

This chapter proposes content-aware privacy protection for LBS to improve privacy protection. This technique creates dummy LBS queries using location–content correlation. This chapter analyzes uncertainty, fake LBS queries, and restored trajectories. Experimental findings show that the suggested strategy works by altering anonymity requirements and semantic relevance ranges.

References

1. Abbasi M, Rezaei H, Menon VG, Qi L, Khosravi MR (2020) Enhancing the performance of flow classification in SDN-based intelligent vehicular networks. IEEE Trans Intell Transp Syst 22(7):4141–4150
2. Mohammadi M, Molaei E, Naserasadi A (2011) A survey on location based services and positioning techniques. Int J Computer Appl 24(5):1–5
3. Xu X, Shen B, Yin X, Khosravi MR, Wan S (2020) Edge server quantification and placement for offloading social media services in industrial cognitive IoV. IEEE Trans Industr Inform 17(4):2910–2918
4. Cao Y, Xiao Y, Xiong L, Bai L, Yoshikawa M (2019) Protecting spatiotemporal event privacy in continuous location-based services. IEEE Trans Knowl Data Eng 33(8):3141–3154
5. Hu C, Li W, Cheng X, Jiguo Y, Wang S, Bie R (2017) A secure and verifiable access control scheme for big data storage in clouds. IEEE Trans Big Data 4(3):341–355

6. Niu B, Li Q, Zhu X, Cao G, Li H (2014) Achieving k-anonymity in privacy-aware location-based services. In: IEEE INFOCOM 2014-IEEE Conference on Computer Communications, pp 754–62
7. Wang L, Yang D, Han X, Wang T, Zhang D, Ma X (2017) Location privacy-preserving task allocation for mobile crowdsensing with differential geo-obfuscation. In: Proceedings of the 26th International Conference on World Wide Web, pp 627–36
8. Li F, Chen Y, Niu B, He Y, Geng K, Cao J (2018) Achieving personalized k-anonymity against long-term observation in location-based services. In: 2018 IEEE Global Communications Conference (GLOBECOM), pp 1–6
9. Forte A, Andalibi N, Greenstadt R (2017) Privacy, anonymity, and perceived risk in open collaboration: a study of tor users and Wikipedians. In: Proceedings of the 2017 ACM Conference on Computer Supported Cooperative Work and Social Computing, pp 1800–1811
10. Zheng X, Cai Z, Li J, Gao H (2017) Location-privacy-aware review publication mechanism for local business service systems. In: IEEE INFOCOM 2017-IEEE Conference on Computer Communications, pp 1–9
11. Huguenin K, Bilogrevic I, Machado JS, Mihaila S, Shokri R, Dacosta I, Hubaux J-P (2017) A predictive model for user motivation and utility implications of privacy-protection mechanisms in location check-ins. IEEE Trans Mob Comput 17(4):760–774
12. Jacob S, Menon V, Al-Turjman FM, VinojP G, Mostarda L (2019) Artificial muscle intelligence system with deep learning for post-stroke assistance and rehabilitation. IEEE Access 7:133463–133473
13. Papernot N, McDaniel P, Sinha A, Wellman MP (2018) Sok: security and privacy in machine learning. In: 2018 IEEE European Symposium on Security and Privacy (EuroS&p), pp 399–414
14. Mohassel P, Zhang Y (2017) Secureml: a system for scalable privacy-preserving machine learning. In: 2017 IEEE Symposium on Security and Privacy (SP), pp 19–38
15. Tavallali P, Yazdi M, Khosravi MR (2019) Robust cascaded skin detector based on AdaBoost. Multimed Tools Appl 78(2):2599–2620
16. Alrahhal MS, Khemakhem M, Jambi K (2017) A survey on privacy of location-based services: classification, inference attacks, and challenges. J Theoret Appl Inf Technol 95(24):6719–6740
17. Liu J, Jiang X, Zhang S, Liu B, Dou W (2022) CPP: a content-aware privacy protection method for location-based service. Expert Syst 39(5):e12907
18. Zheng Y, Xie X, Ma W-Y et al (2010) Geolife: a collaborative social networking service among user, location and trajectory. IEEE Data Eng Bull 33(2):32–39
19. Liu H, Li X, Li H, Ma J, Ma X (2017) Spatiotemporal correlation-aware dummy-based privacy protection scheme for location-based services. In: IEEE INFOCOM 2017-IEEE Conference on Computer Communications, pp 1–9

Chapter 7
Industrial Application Data Security Preservation in Edge Computing

Abstract Authentication is the process of verifying the identity of a user or device attempting to access an application. In an industrial setting, authentication can be used to ensure that only authorized personnel or devices can access critical data. Considered to be a development of cloud computing, edge computing moves computational and storage capabilities to the network edge. Users need to be verified before accessing the geographically dispersed edge servers for several Industrial Internet of Things (IIoT) applications.

In response to this issue, this chapter suggests EdgeAuth, a cutting-edge token-based authentication system that enables quick edge user authentication through cooperation between cloud servers and edge servers. In accordance with the EdgeAuth scheme, edge servers can quickly confirm the login information of users who have already been verified by the cloud server. Users can also be protected by EdgeAuth from a variety of authentication threats, such as the man-in-the-middle assault, replay attack, and denial of service (DoS) attack. By a comparison of latency and throughput against two benchmark systems, the findings of studies carried out in a simulated edge computing environment prove the effectiveness of EdgeAuth.

In an industrial setting, the use of Internet of Things (IoT) devices has become increasingly common. These devices are often connected to a network and transmit data that is critical to the operation of the industrial application. However, with the increased connectivity, the risk of unauthorized access to the application's data has also increased. Therefore, it is important to implement authentication mechanisms to ensure that only authorized users have access to the application's data.

Authentication is the process of verifying the identity of a user or device attempting to access an application. In an industrial setting, authentication can be used to ensure that only authorized personnel or devices can access critical data. This can be achieved through the use of various authentication methods, such as passwords, biometric authentication, or digital certificates.

The importance of implementing authentication in an industrial IoT environment cannot be overstated. Unauthorized access to critical data can result in serious consequences, such as data loss, equipment damage, or even harm to personnel. By implementing authentication mechanisms, organizations can help ensure that

W. Dou et al., *Intelligent Industrial Internet Systems*,
https://doi.org/10.1007/978-981-99-5732-3_7

their data is protected from unauthorized access and maintain the integrity of their industrial processes.

Considered to be a development of cloud computing, edge computing moves computational and storage capabilities to the network edge. Users need to be verified before accessing the geographically dispersed edge servers for several Industrial Internet of Things (IIoT) applications, such as stockpile monitoring and cooperation, Internet of Vehicles, real-time video analysis, etc. It is impossible to duplicate the authentication technique from cloud servers when users are authenticated on edge servers due to the significant delay between the cloud and edge servers and the constrained capacity of edge servers.

In response to this issue, this chapter suggests EdgeAuth, a cutting-edge token-based authentication system that enables quick edge user authentication through cooperation between cloud servers and edge servers. In accordance with the EdgeAuth scheme, edge servers can quickly confirm the login information of users who have already been verified by the cloud server. Users can also be protected by EdgeAuth from a variety of authentication threats, such as the man-in-the-middle assault, replay attack, and DoS attack. By a comparison of latency and throughput against two benchmark systems, the findings of studies carried out in a simulated edge computing environment prove the effectiveness of EdgeAuth.

7.1 Data Security Preservation by Authentication

The primary method for distributing software as a service is now the cloud [1]. It is very easy for customers to access software applications and for app suppliers that offer web application services to install the software. Nevertheless, with the recent expansion of mobile [2] and Internet of Things (IoT) devices [3, 4], a vast array of IIoT applications, including supply-chain management and collaboration, the Internet of Vehicles, real-time video analytics, etc., have posed significant difficulties for cloud computing. Because of their great distance from the centralized cloud servers that host these latency-sensitive IIoT applications [5], users of these apps frequently experience high and unpredictably delays.

By placing edge servers on base stations that are physically close to consumers, the edge computing paradigm offers a potent answer to this problem. These edge servers offer strong IIoT application compatibility. The Internet of Vehicles, for instance, has strict requirements for real-time response [6]. These edge servers can deliver prompt network assistance near to users. Edge servers have a far lower reaction time than cloud servers, making them a more safe computing environment for the Internet of Vehicles. Due to their frequent needs for data contact, real-time video analysis, supply-chain oversight, and collaboration also profit from edge computing's low latency [7]. The main technology for implementing IIoT applications is low-latency edge computing, to put it simply.

Similar to Software-as-a-Service (SaaS) made possible by cloud computing, app suppliers may install apps on edge servers and hire compute and storage resources to run their edge applications. Often, infrastructure service providers like AT&T build up these edge servers. Online businesses like Google will occasionally work with infrastructure service providers to offer edge services. Each edge server connected to a base station serves a certain region. Only when a user is situated inside the coverage area of an edge server may they connect to it wirelessly to access apps hosted by that edge server. The particular coverage restriction is this [5]. Connected users may also delegate computation-intensive activities to nearby edge servers for processing, saving battery life on their mobile devices [8]. Its close proximity to edge servers enables the minimal end-to-end latency that latency-sensitive applications demand.

Since the edge servers are wide geographically spread, adversaries have more chances to cyberattack or illegally access the servers than centrally controlled cloud ones [9]. Users are in danger of cyberattacks if edge servers host apps or data that are incomplete [10]. Unauthorized access jeopardizes the security of the data and programs installed on edge servers in the meantime. An endless stream of authentication assaults at cloud servers has affected several app vendors. Moreover, unauthorized access to edge apps might result in large financial losses [11].

Authentication assaults may be thwarted with proper user authentication. There has been much study on user authentication. From a single password system decades ago to today's multi-factor and even biometric identity authentication, authentication technologies have come a long way [12–14]. These authentication procedures leverage the client–server architecture to have users send their credentials to external cloud servers for verification. In a distributed computer system, such authentication procedures are inappropriate. The user accesses the core network and communicates with faraway cloud servers through a wireless connection to a local base station in the conventional cloud computing architecture. A user's connection to the cloud server remains unchanged even if the base station it's currently connecting to is altered. Instead of interacting with cloud servers, users in an edge computing environment do so with an edge server hosted locally at the base station. Therefore, the user must re-authenticate with the cloud server before accessing the edge application installed on s_j when the user travels from the coverage area of edge server s_i to that of edge server s_j. The large delay introduced by re-authentication is unacceptable in the peripheral computing environment, where low service latency is the fundamental aim. It is straightforward to utilize several peripheral servers rather than a central cloud server for user authentication. The time needed for re-authentication will be shortened as a result. However, peripheral servers have restricted storage capacities due to their compact form factors. This is the upper limit, or capacity. An industrial app provider cannot afford to keep the right credentials of all of its users on each edge server, even if the machines have ample storage capacities.

EdgeAuth, a unique technique that enables speedy peripheral user authentication, is proposed in this chapter to overcome the difficulties above. The idea behind EdgeAuth is to remove the requirement for users to re-authenticate through the cloud server when they move between coverage zones of edge servers by using cloud-edge cooperation. Some of its most important results are:

1. We formally present the model used to investigate the edge user authentication (EUA) problem and describe how the EdgeAuth scheme facilitates fast edge user authentication without cloud server re-authentication.
2. We examine the resistance of EdgeAuth to a variety of common authentication attacks in the edge computing environment, such as the replay attack, DoS attack, and main-in-the-middle attack.
3. We thoroughly validate EdgeAuth's utility in a simulated edge computing environment and compare it to two baseline schemes.

7.2 Edge User Authentication Problem in Industrial Application

Among IIoT consumers, video broadcasting is a representative application. Video service providers offer video streaming services that allow users to view videos. According to the standard cloud computing paradigm, video app providers host their customers' video streaming services on cloud servers. A user u must first register with the app vendor by sending the vendor information about themselves, including user name and password, before being granted access to the app vendor's video streaming service. If it is, then u has successfully authenticated and may watch movies stored in the cloud. The quality of a video streaming service is greatly affected by the ease with which the video may be watched. Figure 7.1 displays the results of our latency tests between the user and both central and remote servers.

The cloud server is an Aliyun node, while the periphery server is a nearby server. Seven days of data testing revealed that the majority of cloud server latency is between 100 and 120 ms and the majority of peripheral server latency is between

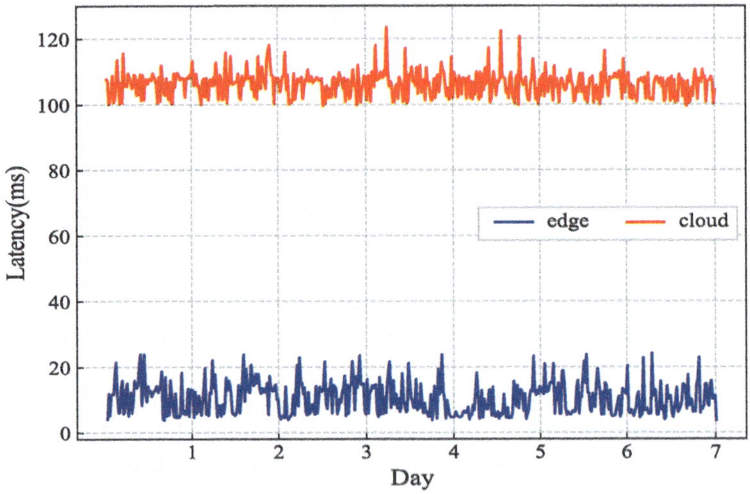

Fig. 7.1 Latency between user and edge/cloud server

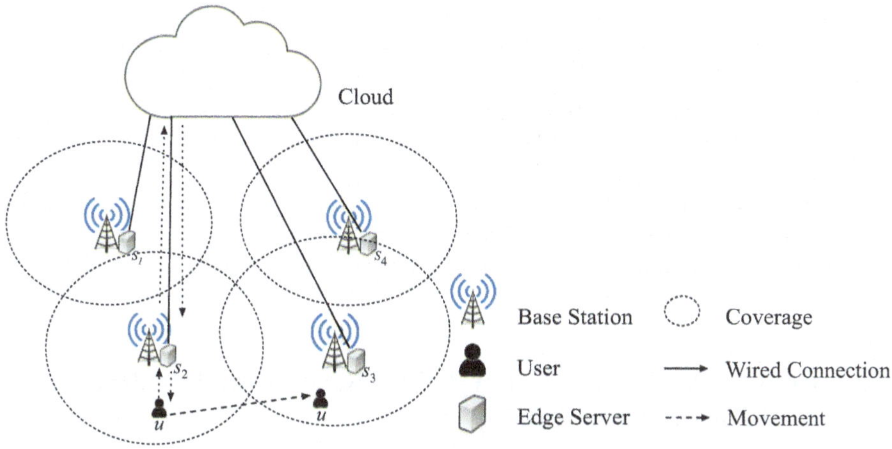

Fig. 7.2 Example edge user authentication scenario

5 and 20 ms. For video streaming services, a latency increase of five to six times between the cloud server and the peripheral server is not insignificant. Therefore, the app vendor may give customers in the area with the best possible service latency by caching the most watched movies in the region in an edge computing environment.

Consider Fig. 7.2, in which the app vendor deploys its services on four edge servers, denoted s_1, ..., s_4, to serve cached videos to users in this region. Each peripheral server is wired to the cloud server and its service coverage is represented by a dashed circle. A valid user u within the coverage area of s_2 desires to access the video application hosted by s_2. It must first demonstrate its identity. It must transmit its username and password to the cloud server for authentication to accomplish this. Assume that user u is authentic and authentication is successful. Next, user u moves from s_2's coverage area to s_3's coverage area while maintaining uninterrupted access to the video service. Without knowledge of u's authentication results, the peripheral application deployed on s_3 is unaware that u has been authenticated. This is because the perimeter server architecture differs from the existing cloud computing architecture. Multiple base stations are connected to the cloud server in actuality. Users must therefore only re-authenticate with their telecom provider when switching base stations. In the edge computing architecture, extremely distributed edge servers are deployed on base stations, and there is no real-time information sharing between edge servers. When transferring base stations, the user must therefore reestablish contact with the edge server.

Sending your credentials to the cloud server for reauthentication is a naive solution. However, u's access to the video application deployed on s_3 will be excessively delayed and potentially interrupted by this solution. The primary issue is the sluggish (and redundant) re-authentication. A possible solution for the app vendor to accelerate the process is to retain all valid user credentials on these peripheral servers. This will permit users, including you, to be swiftly authenticated by peripheral servers as opposed to the cloud server. Unfortunately, storage resources on peripheral servers are limited and costly [15]. On each peripheral server, this solution must retain the correct credentials for all users. Thus, it is not practicable. Moreover, in the highly distributed and dynamic environment of edge computing, edge servers are not always dependable or trustworthy [16]. The accurate credentials of users that are stored on peripheral servers are susceptible to corruption or compromise.

Given the high density of edge servers and their limited coverage areas [17], the key to facilitating fast edge user authentication is to eliminate the need for reauthentication when users cross coverage areas of edge servers.

7.3 EdgeAuth: An Intelligent Token-Based Collaborative Authentication Scheme

We provide the EdgeAuth scheme's system models, general procedure, token architecture, and authentication method in this section.

7.3.1 System Model

There are three primary responsibilities in a scenario involving user authentication at the edge:

1. Cloud Server: The authentication server used by the app provider is located in the cloud. Its primary roles are (i) storing valid user credentials, (ii) authenticating users via a comparison of their login credentials with the valid credentials stored in the database (cloud-side authentication), and (iii) producing and distributing tokens to edge servers performing authentication. The authors of this research make the assumption that cloud servers are secure and reliable because of their centralized administration.
2. Edge Server: Edge Servers are typically deployed at base stations. Under the EdgeAuth protocol, they carry out two primary responsibilities: (i) authenticating users at the network's interface, also known as edge-side authentication; and (ii) collaborating to defend against authentication attacks. In this chapter, it is presumed that edge servers are honest but not entirely reliable [18, 19].
3. User: Users must be authenticated before accessing peripheral applications if they are registered. Replay attacks, denial-of-service attacks, and man-in-the-middle attacks are examples of the types of assaults that adversary users may employ [20].

7.3.2 General Authentication Process

As depicted in Fig. 7.3, the general authentication procedure under the EdgeAuth scheme consists of five distinct phases. The procedure implies the user is already enrolled on the cloud server.

1. To access a service, a user must input their username and password on their end device.
2. The cloud server checks the entered credentials against the user's actual credentials. If the user's credentials are valid, the cloud server will send them an authorized token using a secret key and go on to Step 4. The token has a signature and a time limit that can be checked using the private key. In such a case, the user receives an error notification.
3. When a user's session is about to expire, the cloud server will send a secret key to all of the edge servers it expects the user to visit before the session ends.
4. During a service session, which might include many edge servers, the user gets the token and uses it to access the edge application running on an edge server in close proximity to the user.
5. The token's secret key is compared to the cloud server's secret key to verify the user's identity. Users may access edge apps hosted on edge servers after successfully authenticating.

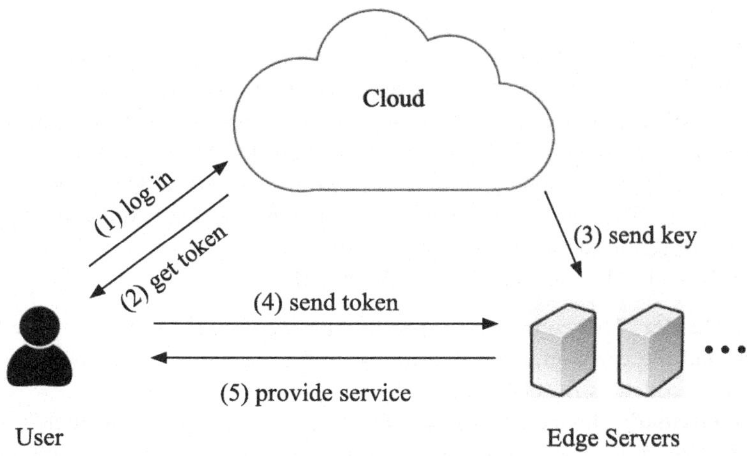

Fig. 7.3 Authentication process under the EdgeAuth scheme

7.3.3 Adversary Model

A trustworthy authentication method is necessary. It has to be able to protect edge servers and the services running on them against threats that may compromise the authentication process. Edge servers in a network in the network's periphery are connected by wiring and may exchange data with one another. Due to the nature of the physical connection between edge servers, we assume that malicious users cannot disrupt this connection or launch a direct assault on the server. However, they may disrupt the wireless channel for malicious purposes or to block legitimate users from reaching the edge servers. In this research, we look at how malicious individuals may attack EdgeAuth using one of the following channels.

1. Replay attack. Attackers may mimic a valid user in an EdgeAuth replay attack by stealing the victim's full communication to the verifier. A hostile user in an edge computing setting might potentially intercept a legal user's communication with an edge server, copy the contents of the message, and retransmit it to either the original edge server or an adjacent edge server.
2. DoS attack. EdgeAuth is vulnerable to denial of service (DoS) attacks, in which malicious users flood edge servers with requests. When an edge server becomes overworked, it stops responding to requests from legitimate users [21].
3. Man-in-the-middle attack. An attacker may perform a man-in-the-middle attack against EdgeAuth by creating a second connection with each side. It has the potential to play back or change transmitted messages between two parties to provide the impression of direct contact. For instance, a man-in-the-middle assault on wireless communication between two parties may be initiated by an inside opponent [22].

7.3.4 Token Structure

When a user needs to get access to a service but doesn't want to hand over personal information like a password, they may use a token instead. One such open standard is OAuth, which allows a user to provide a third-party access to their data across many websites by means of a token. JSON Web Token, or JWT for short, is a popular token standard that facilitates password-less server access for a limited period of time. The JWT is used for authentication in EdgeAuth because of its suitability for use in the network's periphery.

A JSON Web token is composed of three parts:

1. **JWT Header**: This component typically includes metadata regarding the type of token and the cryptographic algorithms used to secure its contents. This string will be encoded as Base64URL.
2. **JWT Payload**: This component contains verifiable security statements, such as the identity of the user and the permitted permissions. Under EdgeAuth, the JWT

payload must include the user's identity and an expiration date. The decoded payload of a JSON Web token is shown below:

{
"uid": "123456789",
"name": "Tom",
"exp": "1640995200"
}

It is important to note that, for simplicity, the expiration period is represented by a specific time, which should correspond to the program's timestamp.

3. **JWT Signature**: This component is used to guarantee the token's integrity, i.e., that it has not been tampered with or fabricated. It is generated by using the algorithm specified in the header to hash the header, the payload, and a unique secret.

7.3.5 *Registration*

To access an edge application, a user must have enrolled on the cloud server of the app vendor by supplying the necessary information. Following are the procedure:

1. The user enters a username (id) and a password (pw), and computes a A = hash (id || pw), where hash() is a hash function, e.g., SHA256. It then saves A in its local storage.
2. The user sends id, hash(pw) to the cloud server.
3. The cloud server checks whether id is already used. If so, it rejects the registration request. Otherwise, it stores id and hash(pw), and sends the app vendor's public key to the user.

7.3.6 *Cloud-Side Authentication*

There are numerous authentication schemes that use various methods to verify user identity, such as username/password, biometrics, and so on. All of these methods can be used to facilitate cloud-side authentication with the EdgeAuth protocol. In this chapter, it is assumed that the cloud server verifies a user's credentials based on their username and password.

Figure 7.4 depicts the cloud-side authentication and edge-side authentication processes for the EdgeAuth scheme. Before accessing an edge application, a user u must opt into the cloud server with a username and password and obtain a token. The method is as follows:

1. User u enters his username and password on its end-device.
2. It computes $A^{'}$ = hash(id || pw), and checks whether $A^{'} = A$ to perform a local authentication process. If yes, go to the next step. Otherwise, return to Step 1.

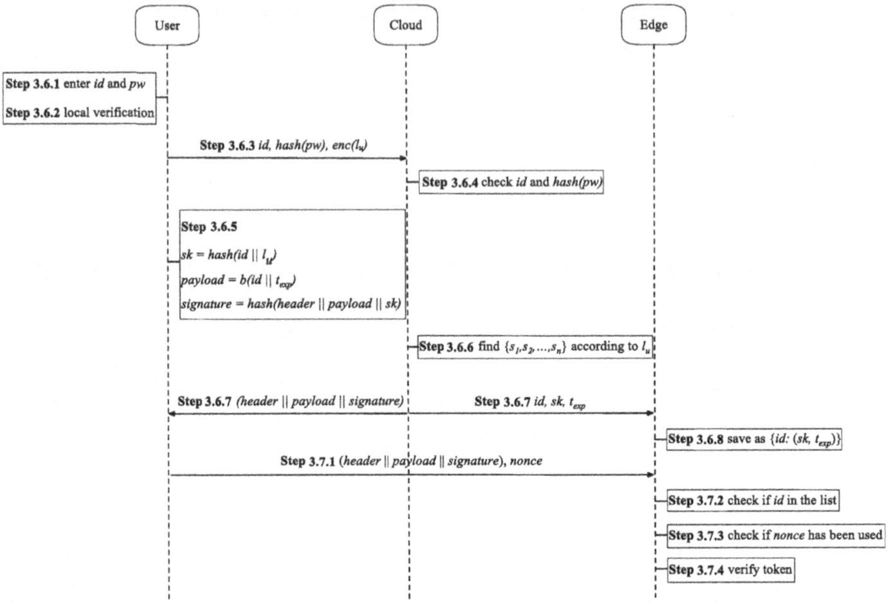

Fig. 7.4 Processes for login and authentication under the EdgeAuth scheme

3. User u sends id, hash(pw), along with the user's encrypted location information $el_u = (l_u)^e \bmod p$, to the cloud server.
4. The cloud server checks if id and hash(pw) are both correct. If not, the cloud server sends a fail message to user and the user needs to retype and resubmit its id and hash(pw). Otherwise, go to the next step.
5. The cloud server computes $l_u^{'} = (el_u)^d \bmod p$ and generates a secret $sk =$ hash (id $\|$ l_u) and sets the expiration time t_{exp}, then computes payload $= b$(id $\|$ t_{exp}), where b() is a binary coding function that converts an object to a string representation . Next, it generates the header in the JWT and computes signature $=$ hmac (header $\|$ payload $\|$ sk).
6. Based on user u's location l_u, the cloud server predicts the edge servers that user u may need to access before the expiration time, denoted as $\{s_1, s_2, \cdots, s_n\}$. EdgeAuth is not dependent on any specific prediction methods. For example, user u's trajectory may be predicted to identify the edge servers covering its route before the expiration time. A radius r_u may be predicted that depicts a *potential movement area* with l_u as the center that covers user u's movement before the expiration time. The prediction of user movement or user trajectory is a research topic orthogonal to our study. In this chapter, we assume that a potential movement area is predicted and user u's *potential edge servers* within the potential movement area are identified, i.e., those that may be accessed by user u before the expiration time. An example can be found in Fig. 7.5.

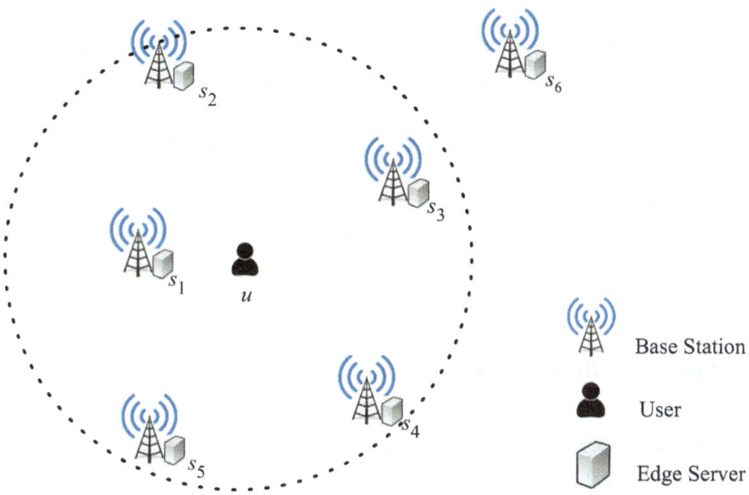

Fig. 7.5 User u's potential movement area

7. The cloud server sends a token {*header*, *payload*, *signature*} and a random large prime number x to user u. It also sends user u's id, its corresponding secret sk, the large prime number x and expiration time t_{exp} to edge servers $\{s_1, s_2, \cdots, s_n\}$.
8. Edge servers $\{s_1, s_2, \cdots, s_n\}$ save the received user u's *id*, *sk*,*x* and t_{exp} to their lists of potential users. It will stay on the list for edge-side authentication before t_{exp}.

7.3.7 Edge-Side Authentication

User u must obtain the token from the cloud server before accessing any peripheral servers. Every time it attempts to access an edge server, you will send the edge server the received token and a randomly generated nonce for authentication. In this study, we presume that every potential edge server hosts the intended edge application for user u. If user u's desired edge application is not hosted by any of its adjacent edge servers, the user will bypass edge-side authentication and access the cloud-hosted application.

Upon the receipt of user u's token, edge server s can authenticate user u. This edge-side authentication process is as follows:

1. User u randomly generates a large prime number y and computes a nonce $= xy$. Then the user sends an authentication request that contains a token {header, payload, signature} and the nonce to edge server s.
2. Edge server s extracts *id* from the token and checks whether it can find in its list of potential users. If so, go to the next step. Otherwise, authentication fails, and user authentication process turn back to step 3 in Sect. 7.3.6.

3. Edge server s computes $y' = \text{nonce}/x$ and judge whether y' is an integer. If so, s checks whether nonce matches user u's nonce recorded in its list of potential users. If not, it records nonce in the list and goes to the next step. Otherwise, authentication fails.
4. The edge server s compares the token of the user u with the corresponding secret sk in the list of potential users. If it is a match, authentication is successful and the periphery server will commence providing service. In the event that the authentication fails, the user must return to the cloud-side authentication phase and resume the authentication procedure.

7.3.8 Nonce Synchronization

As stated in Sect. 7.3.7, every time a user u attempts to access an edge server, it will generate a random nonce for authentication. Using a random nonce ensures that a token generated by the cloud server for user u can only be utilized by user u. The replay assault is a possible hazard to EdgeAuth. To launch a replay attack, an adversary intercepts the entire authentication request sent by a user u to an edge server s and sends it to s or other edge servers in an attempt to circumvent authentication. By including a nonce in u's authentication request that it transmits to s, EdgeAuth ensures that an adversary cannot utilize this authentication request to cheat authentication. As described in Step 3 of Sect. 7.3.7, the edge server s will determine if the nonce included in user u's authentication request has previously been used by a user with the same id. If so, the edge server recognizes this as a replay attack.

However, an adversary may still initiate a replay attack against other edge servers that have never authenticated user u, as these edge servers do not include user u's nonce in their own lists of potential users. The key to mitigating this attack is to ensure that all potential peripheral servers are aware of the use of any nonce by an authenticated user u. Consequently, under the EdgeAuth scheme, when an edge server effectively authenticates a user based on a nonce, it alerts other prospective edge servers to the use of this nonce.

After recognizing the freshness of the nonce, edge server s will add the nonce to the list and broadcast a message containing id and nonce to edge servers within $2r_u$ radius to ensure that $\{s_1, s_2, \cdots, s_m\}$ can receive the message. For each edge server receiving the message, it checks whether id is stored in its list of potential users and adds nonce to the list of the corresponding user if so. Otherwise, it ignores the message.

Through nonce synchronization, every potential edge server will be able to identify any replay attack with this nonce at the expense of minimal synchronization costs. It serves as the foundation for EdgeAuth's defense against numerous cyberattacks.

7.4 Performance and Security Analysis of EdgeAuth

In this section, we experimentally compare EdgeAuth to two representative periphery user authentication schemes based on traditional cloud-based schemes.

7.4.1 Research Questions

EdgeAuth enables rapid user authentication in the edge computing environment by enabling authorized users to access edge servers without repeated cloud server authorization. In this manner, EdgeAuth ensures that authorized users receive uninterrupted service when moving from one edge server's coverage area to another. This is dependent on two variables: (1) the user's potential movement range after authorization by the cloud server; and (2) the user's movement pace. Consequently, the investigations will provide responses to the following two questions:

1. **RQ 1**: How does the movement range of a user after initial authentication affect EdgeAuth performance?
2. **RQ 2**: Is EdgeAuth capable of rapidly authenticating a user who is moving at varying speeds?

7.4.2 Experiment Setup

To facilitate the experiments, we simulate a 5G-powered Melbourne Central Business District. A total of 125 virtual machines are deployed in a private data center to simulate 125 edge servers whose physical locations are extracted from the widely used public EUA dataset. To simulate the cloud server, an Amazon EC2 instance is contracted. The simulation of 817 mobile users whose locations are extracted from the EUA dataset. Experiments indicate that it takes 5–20 ms to transmit a message between a user and an edge server, 80–100 ms between an edge server and a cloud server, and 100–120 ms between a user and a cloud server.

EdgeAuth does not rely on specific methodologies for predicting a user's prospective edge servers, as described in Sect. 7.3.6. In each investigation, 100 participants are selected at random from the 817 participants. After authorization by the cloud server, each user *u* moves within a circular area with its origin as the center and a radius ranging from 500 to 1000 meters in increments of 100. To answer the second question, we simulate *u*'s motion at low, medium, and high travel rates. The top 50, 100, and 200 users closest to you are identified from the remaining 716 (816–100) users to simulate your trajectory. Specifically, commencing from its initial location, *u* will always travel to the 50, 100, or 200 location that is the next closest. In this manner, we conduct three sets of experiments: Sets 1, 2, and 3. Every time user *u* travels to a new location, it will attempt to connect to the nearest edge server. Following this pattern, 100 users will simultaneously move throughout the simulated area in each experiment. They are authenticated using three rival authentication schemes. Each user will continue to move until they have accessed 20 perimeter servers, requiring 20 authentications. Depending on the authentication schemes, the cloud server or peripheral servers may conduct these authentications. We calculate the total latency incurred by each user's 20 authentications using various schemes. Finally, we report the average total latency incurred by 100 users across their 20 authentications, as well as the authentication throughput.

7.4.3 Baseline Schemes

EdgeAuth, unlike other cryptographic protocols designed for cloud or edge computing environments, attempts to actualize edge authentication from a different perspective than optimizing existing schemes to reduce latency. In the experiments, EdgeAuth is contrasted to two baseline authentication schemes.:

1. **Cre2C**: Cre2C stands for cloud credentials. This technique represents a direct copy of the cloud computing mode authentication method to the periphery computing mode. Thus, the cloud server stores the registration credentials of the user. There are no user credentials on the edge server, so the user's logon information must be sent to the cloud server for verification when the user attempts to log in. Under this scheme, whenever a user attempts to access an application deployed on a periphery server, the cloud server will be contacted for authentication.
2. **Cre2E**: Cre2E is an abbreviation for credentials to edge. This method, unlike Cre2C, caches some user credentials on the periphery server. Under this scheme, when a user attempts to access an application deployed on an edge server, the edge server examines if the user's credentials are stored in its cache. If so, authentication is successful. If not, the edge server retrieves this user's credentials from the cloud and caches them for authentication before the expiration time. When a user returns to the coverage area of this edge server, it will be authenticated directly by the edge server, bypassing the cloud server.

During the authentication procedure, Cre2C, Cre2E, and EdgeAuth employ the same RSA encryption algorithm and SHA256 hash function in the experiments.

7.4.4 Performance Metrics

It requires time to accomplish an individual's authentication. As discussed in Sect. 7.1, one of the primary goals of the edge computing paradigm is to reduce latency, and the primary design goal of EdgeAuth is to provide quick authentication. In the experiments, the average authentication latency of users is used to evaluate the performance of EdgeAuth. It is calculated as follows and indicates the authentication performance from the user's perspective:

$$T_{\text{auth}} = T_{\text{comp}} + T_{\text{tran}}$$

where T_{comp} is the time taken for a cloud server or an edge server to complete the computation for authentication, e.g., password verification, token verification, etc., and T_{tran} is the time for message transmission. In the investigations, we report the average total latency incurred by 20 authentications during their access to 20 peripheral servers during their movements.

As discussed in Sect. 7.1, peripheral servers have limited resources, in contrast to cloud servers. Their assets are more costly than those of cloud servers. Thus, user authentication must be efficient and incur minimal hiring costs for peripheral server resources. To evaluate the efficacy of EdgeAuth from this vantage point, we measure its authentication throughout each investigation, i.e., the number of user authentications it completes per second. As discussed previously in Sect. 7.4.2, all 100 users move simultaneously in each experiment. Consequently, with a total of 2000 (20×100) authentications in each experiment, the authentication throughput is dependent on the user who requires the most time to authenticate overall. A high throughput indicates a high level of authentication efficacy and minimal resource costs. It indicates the authentication efficacy from the perspective of the app vendor. To eradicate the impact of resource disparity between edge servers, in the experiments, individual edge servers authenticate users within their respective coverage areas sequentially.

Edge-side authentication under the EdgeAuth scheme incurs storage and communication overheads due to the storage of authentication information on edge servers and the synchronization of nonce values among edge servers. Both variables correlate positively with the number of prospective edge servers within a user's potential movement area. To evaluate the costs associated with EdgeAuth, we determine the average number of potential edge servers per user.

7.4.5 Experimental Results

Figures 7.6, 7.7, and 7.8 depict the results of Sets 1 through 3 of an experiment. EdgeAuth can authenticate users with minimal authentication latency and high throughput. The 20 authentications for each user are completed in an average of 739.33 milliseconds, or 36.97 milliseconds per authentication, which is 460.49% and 346.26% faster, respectively, than Cre2C and Cre2E. EdgeAuth obtains a throughput of 32.37 authentications per second across Sets #1 through #3, which is 445.62 and 252.32% higher than Cre2C and Cre2E, respectively. The aggregate results demonstrate the significant benefits of authenticating users at the edge as opposed to in the cloud. The numbers also indicate that the travel velocities and

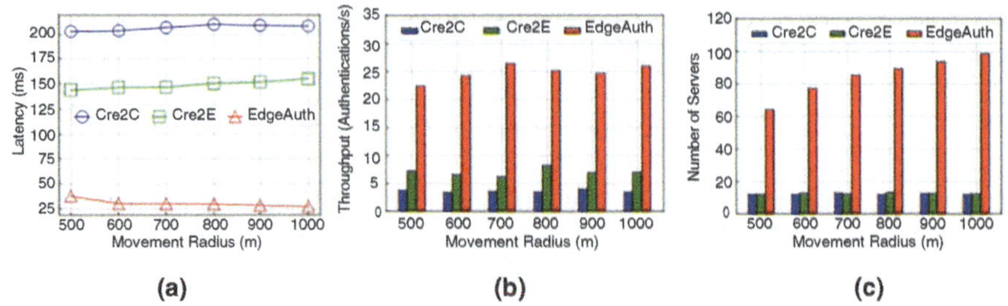

Fig. 7.6 Performance with users moving at low speed (Set #1). (a) Latency vs. r_u . (b) Throughput vs. r_u . (c) Number of servers vs. r_u

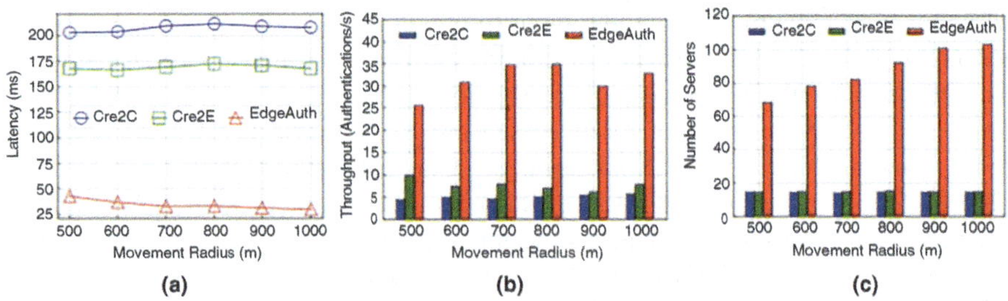

Fig. 7.7 Performance with users moving at medium speed (Set #2). (**a**) Latency vs. r_u (**b**) Throughput vs. r_u . (**c**) Number of servers vs. r_u

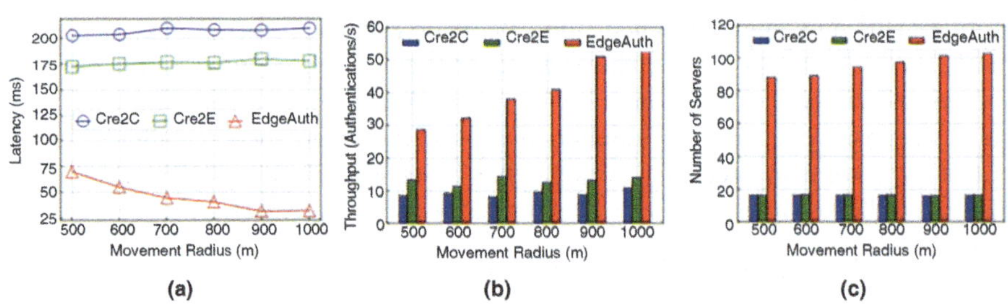

Fig. 7.8 Performance with users moving at high speed (Set #3). (**a**) Latency vs. r_u . (**b**) Throughput vs. r_u . (c) Number of Servers vs. r_u

movement radius of users have no effect on the efficacy of the Cre2C and Cre2E schemes. Therefore, the following discussion will concentrate on the efficacy of EdgeAuth.

Figure 7.6 depicts the results of Set 1 when users travel at modest velocities. As shown in Fig.7.6a, the authentication latency decreases as the movement radius (r_u) increases from 500 meters to 600 meters under the EdgeAuth scheme. With r_u=500, a user's prospective movement area is modest, and even at a low travel speed, the user can move outside of this area. Before it can be authenticated at the edge in a new potential movement area, the user will need to be re-authenticated by the cloud server according to the procedure described in Sect. 7.3.6. These cloud-based authentications increase the average authentication delay for consumers. As r_u approaches 600 meters, the prospective movement area is sufficiently large to encompass users' trajectories, allowing the majority of EdgeAuth authentications to be performed by edge servers with minimal latency. As r_u increases, the authentication latency continues to decrease, but only marginally. With the majority of users authenticated at the edge, the authentication system, including the edge servers and the cloud server, achieves a high throughput under the EdgeAuth scheme, as shown in Fig. 7.6b, outperforming Cre2C and Cre2E in Set #1 by 578.77% and 254.5%, respectively. As described in Sect. 7.4.4, the authentication throughput is dependent on the user that requires the most time to authenticate with EdgeAuth 20 times, i.e., the duration of the experiment. As shown in Fig. 7.6b, although an increase in r_u can reduce the average authentication latency of users, it does not substantially increase system throughput. Figure 7.6c depicts the average number of potential peripheral servers per user. The correlation between its increase and r_u is not unexpected. This is the cost of enabling edge-side authentications via the EdgeAuth scheme.

Figure 7.7 illustrates the results of Set 2 when users travel at moderate velocities. In this series of investigations, the prospective movement areas of users can still encompass the majority of user trajectories. Thus, the performance advantages of EdgeAuth over the Cre2C and Cre2E schemes are comparable to those in Set 1. The increase in its operating expenses is also comparable to Set 1.

Figure 7.8 depicts the results of Set 3 when users travel at high velocities. In this set of investigations, users frequently leave their permitted movement areas, necessitating re-authentication by the cloud server to facilitate edge-side authentication for access to edge servers. This causes additional authentication latency and increases the average authentication latency experienced by users. As shown in Fig. 7.8a, it is notably evident when r_u is modest. In this set of investigations, EdgeAuth can reduce the average authentication latency of users to as little as 70.12 milliseconds with 88 edge servers ($r_u = 1000$ in Fig. 7.8c). In this series of experiments, users are traveling at high speeds, which can easily lengthen the duration of the experiment if their trajectories do not fall within the prospective movement areas. Consequently, a significant decrease in authentication latency accompanied by a large r_u can effectively increase authentication throughput. In Fig. 7.8b, the authentication capacity attained by EdgeAuth increases from 28.71 to 55.06 authentications per second as r_u increases from 500 to 1000 meters.

7.4.6 Discussion

The experimental results demonstrate that user mobility has a significant impact on EdgeAuth's performance. A user moving at a high pace is more likely to depart the collective coverage area of its prospective edge servers, necessitating reauthorization by the cloud server. This highlights the significance of a precise user trajectory prediction. If the user's movement trajectory can be accurately predicted, EdgeAuth will send secrets to edge servers based on the user's trajectory, obtaining the highest efficacy at the lowest cost. In other words, an accurate user trajectory prediction method can considerably enhance the efficacy of EdgeAuth. In fact, user trajectory prediction is a prominent topic of research independent of EdgeAuth. We did not employ a sophisticated method for predicting potential peripheral servers for this reason. Instead, we opted for the simplest solution, which consisted of a circular active area representing the user's range of motion and a saturated secret distribution encompassing all the servers that might interact. This approach has the advantage of minimizing the dependence of EdgeAuth's performance on the prediction of user movement trajectory and directly analyzing the method's performance. Thus, the results presented in Figs. 7.6, 7.7, and 7.8 demonstrate the performance minimum for EdgeAuth.

Under the EdgeAuth protocol, as described in Sect. 7.4.4, the nonce value must be synchronized between edge servers. Communication costs are proportional to the number of prospective servers, which is unlikely to be very significant. In addition, communication costs are incurred only at the network's interface and not over the backhaul network. This is consistent with the core objective of peripheral computing. EdgeAuth imposes storage costs on peripheral servers. As described in Sect. 7.3.7, each edge server must maintain an inventory of potential users, or users who may migrate into its coverage area. These users are from neighboring regions. Therefore, it is unlikely that the number of potential users on the list is excessively large. In addition, each potential user on the list has an expiration date. Thus, the list will not expand indefinitely and the peripheral servers will not incur significant storage costs.

There is a substantial amount of research devoted to mitigating various authentication assaults in the cloud computing environment. The edge computing environment is fundamentally distinct from the cloud computing environment due to a number of distinguishing features, such as the coverage constraint and capacity constraint discussed in Sect. 7.1. Conventional cloud computing authentication assaults will adapt to the edge computing environment, acquiring new characteristics and posing new threats. In contrast to some cryptographic analyses of security, our scheme places less emphasis on security than these protocols. In this section, we discuss how three such authentication attacks, including the replay attack, DoS attack, and man-in-the-middle attack, are mitigated by the EdgeAuth scheme in the edge computing environment.

7.4.7 Security Analysis

7.4.7.1 Defense Against Replay Attack

The key to preventing a replication attack is the uniqueness of the authentication information used by each user when communicating with the server. Typically, timestamps or randomly generated nonce values are used as part of the authentication information in the cloud computing environment to prevent replay attacks. This allows the server (as the authenticator) to examine the authentication information's uniqueness. Timestamps frequently require global time synchronization and are difficult to implement. Consequently, nonce-based authentication has been thoroughly studied and implemented. In general, a unique nonce value is generated at random for each server service session. By examining this nonce value, the authenticator can determine whether a token has been previously used by another user. Consider a legitimate user u_l with a nonce n_l and an adversarial user u_m with a nonce n_m attempting to impersonate u_l by using a message intercepted from u_l. The server will discover that u_l has used the token provided by u_m with the nonce n_l. Thus, it will be able to determine that n_m is invalid.

Because all of an app's users are typically authenticated by a single cloud server, once-based authentication schemes can be readily employed in the cloud computing environment. In the edge computing environment, however, a user's service session may entail multiple geographically dispersed edge servers. It is challenging for these edge servers to ensure that user nonce values are globally distinct. Under the EdgeAuth scheme, a user u's potential edge servers, i.e., those within u's potential movement area that may be accessed by user u, communicate with each other to synchronize user u's login status to prevent replay attacks. This ensures that user u can only access a single perimeter server at any given time. When u logs into s_1 with nonce n_0, for instance, an adversary replays the message to s_2. Then s_2 will recognize that n_0 exists in its cache. EdgeAuth can thus defend against replay attacks by identifying adversarial users attempting to impersonate other users with intercepted messages.

7.4.7.2 Defense Against DoS Attack

DoS (denial-of-service) assaults are one of the most common types of cyberattacks in the cloud environment, and they cause significant annual economic losses. The fundamental concept is to bombard the target server with excessive attack requests to deplete its processing, memory, and/or bandwidth resources, rendering it incapable of handling legitimate requests. To facilitate and regulate increasing workloads, app vendors frequently deploy their applications on multiple (virtual) servers in the cloud environment. A server that is under attack may transfer its inbound duties to other servers. Unfortunately, this elasticity will forward assault requests to other servers, allowing the attack to engage and overwhelm the entire system.

Edge servers are especially susceptible to DoS assaults due to their severely limited resources in comparison to cloud servers. Similar to cloud servers, they can transfer user queries to each other to mitigate DoS attacks. EdgeAuth provides three levels of protection in addition to existing mechanisms against DoS attacks in the edge computing environment. First, EdgeAuth restricts access to peripheral applications to only authenticated users. Second, adversarial users who have intercepted u's communications can only access u's potential peripheral servers.

They cannot therefore target other peripheral servers. Third, adversarial users cannot assume the identity of authenticated users. Consequently, they are unable to coordinate DoS attacks against potential peripheral servers.

If the attack requests originate from multiple sources, the DoS attack transforms into a DDoS (distributed denial-of-service) attack, which is more difficult to mitigate. Highly distributed edge server deployment in the edge computing environment offers DDoS attacks more opportunities to perform. An adversary can attempt multiple logins on multiple peripheral servers. However, as stated previously, the edge server can rapidly eradicate some false requests by verifying the user id in the token, which also provides some DDoS resistance.

7.4.7.3 Defense Against Man-in-the-Middle Attack

Authentication is a promising method for preventing man-in-the-middle attacks because it restricts server communication to only authenticated users. Under the EdgeAuth protocol, cloud-issued tokens are required to access peripheral applications. Without a token, an adversary cannot fool a server at the network's periphery into believing it is a legitimate user. This instantaneously renders a man-in-the-middle attack impossible to succeed.

7.5 Summary

App providers are able to deploy their apps on edge servers in a computing environment that uses edge computing, which provides customers with a low service latency. This opens up a wide variety of fresh potential for app producers while simultaneously presenting them with a number of unexpected problems.

EdgeAuth is the name of our unique token-based approach that we suggested in this chapter to enable quick user authentication in edge computing. This is the very first attempt made to solve the challenge of authenticating users at the edge. When a user wishes to access an app vendor's edge apps that are deployed on edge servers, it must first be authorized by the app vendor's cloud server to comply with the requirements of the EdgeAuth scheme. After that, the authentication of the device is handled by the edge servers that are located nearby whenever it enters the coverage area of an edge server. EdgeAuth is able to protect protection against a number of the conventional authentication threats that might occur in an edge computing environment. The utility of this approach is demonstrated by the findings of tests carried out on a real-world dataset that is put to widespread usage, and the findings are compared with two typical edge user authentication systems.

An edge server will, in accordance with what was covered in Sect. 7.3.7, remove a user u from its list of possible users when the user's secret becomes invalid. There is a problem with the time being shown inconsistently between the edge server and the user interface. If the secret has expired on the clock of the edge server but not on *u*'s clock, then *u* will need to renew its token with the cloud server a little bit sooner than usual. We will make an effort to mitigate the effects that this problem has on EdgeAuth.

References

1. Sandhu AK (2021) Big data with cloud computing: discussions and challenges. Big Data Min Analyt 5(1):32–40
2. Qi L, Lin W, Zhang X, Dou W, Xialong X, Chen J (2023) A correlation graph based approach for personalized and compatible web APIs recommendation in mobile APP development. IEEE Trans Knowl Data Eng 35(6):5444–5457
3. Li J, Sai AMVV, Cheng X, Cheng W, Tian Z, Li Y (2020) Sampling-based approximate skyline query in sensor equipped IoT networks. Tsinghua Sci Technol 26(2):219–229
4. Mabrouki J, Azrour M, Dhiba D, Farhaoui Y, El Hajjaji S (2021) IoT-based data logger for weather monitoring using Arduino-based wireless sensor networks with remote graphical application and alerts. Big Data Mining and Analytics 4(1):25–32
5. He Q, Cui G, Zhang X, Chen F, Deng S, Jin H, Li Y, Yang Y (2019) A game-theoretical approach for user allocation in edge computing environment. IEEE Trans Parallel Distributed Syst 31(3):515–529
6. Xu X, Jiang Q, Zhang P, Cao X, Khosravi MR, Alex LT, Qi L, Dou W (2022) Game theory for distributed IoV task offloading with fuzzy neural network in edge computing. IEEE Trans Fuzzy Syst 30(11):4593–4604
7. Qi L, Chunhua H, Zhang X, Khosravi MR, Sharma S, Pang S, Wang T (2020) Privacy-aware data fusion and prediction with spatial-temporal context for smart city industrial environment. IEEE Trans Industr Inform 17(6):4159–4167
8. Ren J, Li J, Liu H, Qin T (2021) Task offloading strategy with emergency handling and Blockchain security in SDN-empowered and fog-assisted healthcare IoT. Tsinghua Sci Technol 27(4):760–776
9. Li B, He Q, Chen F, Jin H, Xiang Y, Yang Y (2020) Auditing cache data integrity in the edge computing environment. IEEE Trans Parallel Distributed Syst 32(5):1210–1223
10. Li B, He Q, Chen F, Jin H, Xiang Y, Yang Y (2021) Inspecting edge data integrity with aggregated signature in distributed edge computing environment. IEEE Trans Cloud Computng 10(4):2691–2703
11. Al-Rubaie M, Morris J, Chang. (2016) Reconstruction attacks against mobile-based continuous authentication systems in the cloud. IEEE Trans Inf Forensics Secur 11(12):2648–2663
12. Yang Y, Deng RH, Bao F (2006) A practical password-based two-server authentication and key exchange system. IEEE Trans Depend Sec Computng 3(2):105–114
13. Choudhury AJ, Kumar P, Sain M, Lim H, Jae-Lee H (2011) A strong user authentication framework for cloud computing. In: 2011 IEEE Asia-Pacific Services Computing Conference, pp 110–15
14. Odelu V, Das AK, Goswami A (2015) A secure biometrics-based multi-server authentication protocol using smart cards. IEEE Trans Inf Forensics Secur 10(9):1953–1966
15. Xu X, Li H, Weijie X, Liu Z, Yao L, Dai F (2021) Artificial intelligence for edge service optimization in internet of vehicles: a survey. Tsinghua Sci Technol 27(2):270–287
16. Li B, He Q, Cui G, Xia X, Chen F, Jin H, Yang Y (2020) READ: robustness-oriented edge application deployment in edge computing environment. IEEE Trans Serv Comput 15(3): 1746–1759
17. Tong Z, Ye F, Yan M, Liu H, Basodi S (2021) A survey on algorithms for intelligent computing and smart city applications. Big Data Min Analyt 4(3):155–172

18. Sun H-M, Chen Y-H, Lin Y-H (2011) oPass: a user authentication protocol resistant to password stealing and password reuse attacks. IEEE Trans Inf Forensics Secur 7(2):651–663
19. Zhou L, Varadharajan V, Hitchens M (2015) Trust enhanced cryptographic role-based access control for secure cloud data storage. IEEE Trans Inf Forensics Secur 10(11):2381–2395
20. Jiang X, Dou R, He Q, Dou W (2023) EdgeAuth: An intelligent token-based collaborative authentication scheme. Software: Practice and Experience, Early Access. https://doi.org/10.1002/spe.3206
21. He Q, Wang C, Cui G, Li B, Zhou R, Zhou Q, Xiang Y, Jin H, Yang Y (2021) A game-theoretical approach for mitigating edge DDoS attack. IEEE Trans Depend Sec Comput 19(4):2333–2348
22. Vondráček M, Pluskal J, Ryšavý O (2018) Automated man-in-the-middle attack against Wi-fi networks. J Digit Forensic Secur Law 13(1):9

Chapter 8
Attribute-Based Intelligent Data Management for Industrial Application

Abstract In the area of industrial engineering, a product is designed and refined, step by step, from different respects promoted by various autonomous roles, such as design engineers, process engineers, manufacturing engineers, maintenance engineers, product users, etc. As a result, the dataset of a product is often generated and shared in a cross-platform way. It is a typical distributed product evolutionary process associated with various data versions. It often challenges the consistency of product data management (PDM) that accompanies a product's lifecycle. With this scenario, an effective integrated environment or platform is indispensable for harmonizing various product data's versions produced at different stages. In view of this challenge, a blockchain-driven method for product data's version management in the industrial metaverse is investigated in this chapter. In our method, an industrial metaverse is used to coordinate the various roles for strengthening the aggregation among more engineering scenes in a virtual way, and a smart contract of blockchain is used to guarantee the consistency of the data version. Concretely speaking, multi-authority attribute-based encryption is adopted to ensure the security of shared version data. Moreover, blockchain is taken advantage of to provide a platform for data sharing under the execution of several smart contracts. Consortium blockchain is deployed in an experimental environment, which validates the performance of our method. At last, the version management method presented in this chapter has been deployed in a real product data management system.

In the area of industrial engineering, a product is designed and refined, step by step, from different respects promoted by various autonomous roles, such as design engineers, process engineers, manufacturing engineers, maintenance engineers, product users, etc. As a result, the dataset of a product is often generated and shared in a cross-platform way. It is a typical distributed product evolutionary process associated with various data versions. It often challenges the consistency of product data management (PDM) that accompanies a product's lifecycle. With this scenario, an effective integrated environment or platform is indispensable for harmonizing various product data's versions produced at different stages. In view of this challenge, a blockchain-driven method for product data's version management in industrial metaverse is investigated in this chapter. In our method, an industrial metaverse

W. Dou et al., *Intelligent Industrial Internet Systems*,
https://doi.org/10.1007/978-981-99-5732-3_8

is used to coordinate the various roles for strengthening the aggregation among more engineering scenes in a virtual way, and a smart contract of blockchain is used to guarantee the consistency of the data version. Concretely speaking, multi-authority attribute-based encryption is adopted to ensure the security of shared version data. Moreover, blockchain is taken advantage of to provide a platform for data sharing under the execution of several smart contracts. Consortium blockchain is deployed in an experimental environment, which validates the performance of our method. Finally, the version management method presented in this chapter has been deployed in a real product data management system.

8.1 Product Data Management in Industrial Application

Product data management (PDM) is a technology that uses software or other tools to track and control data related to a particular product [1]. The data tracked helps the business connect and communicate product data between scattered teams. PDM is part of product lifecycle management and acts as a central repository of processes and product history, in which each model, design drawing, or document is kept in. PDM also facilitates integration and data exchange among all business roles that interact with the product [2].

In practice, a product is designed and refined by various autonomous roles, such as design engineers, process engineers, manufacturing engineers, maintenance engineers, product users, etc. Moreover, the product design process is always phased, iterative, and nonlinear. Design always follows a certain workflow that each stage must be proofread, reviewed, and approved by the management process, and finally form a formal version of the release. Throughout the process, design objects undergo many modifications and state changes [3]. This process forms different version data and spawns version management, which is a part of PDM. Design engineers always want to have access to the previous version data of the design object at any time, and they also want to keep the intermediate results of the design process constantly improved, so that they can go back to the previous version in the future if the current design is not reasonable or the results do not meet the requirements. In practice, however, the various roles involved in PDM are often in different departments and cannot communicate in real-time. As a result, the dataset of a product, especially the version data, is often generated and shared in a cross-platform way. In real production, such cross-platform communication seriously impedes the efficiency of version data sharing and management.

The metaverse provides a novel solution for cross-platform data sharing with real-time communication requirements in real life. The concept of the metaverse first appeared in the 1992 science fiction novel Avalanche, which depicts a vast virtual reality world. The concept has since been repeated in games and movies where humans can work, play, and interact through avatars [4]. In recent years, with the development of the Internet, communication, and artificial intelligence technologies, the discussion of the metaverse has gradually increased [5]. The concept of metaverse has become even more popular, especially with the COVID-19 pandemic and the growing need for online social networking [6]. The growing popularity of the metaverse is inseparable from the improvement of infrastructure and breakthroughs in key technologies [7]. For instance, the development of cloud computing and edge computing platforms, as well as the integration of blockchain applications such as trusted authentication and smart contracts, provide good platform support for the realization of the metaverse [8, 9]. However, the technological forces that make up the concept of the metaverse are not yet complete and its development is still in the preliminary stage. More importantly, there is no unified consensus on the concept of the metaverse, which lacks a concrete landing scene.

Mapping the physical and virtual worlds is not a new technology. Digital twin maps physical entities to the virtual world and builds models. The collected data is analyzed and the model is simulated in real time to monitor and optimize the physical entities [10]. Digital twinning is widely used in industry because virtualization models are easier to test and validate during product development. In production, digital twinning technology [11] can simulate the operation of equipment and parameter adjustment changes. The application of metaverse-related technologies in the industrial field will empower various scenarios of the lifecycle of industrial products. Thus, industrial metaverse is a feasible application of metaverse with a realistic foundation and enormous potential.

As the application of metaverse-related technologies in the industrial field, the industrial metaverse fully deploys the design, process, manufacturing, operation, maintenance, and other links and scenes in the real industrial environment in the virtual space. In the industrial metaverse, all roles are mapped into the virtual world, and the physical distance between them completely disappears. These roles interacting with products can share data more easily, but data is also more accessible to people who are not involved [12]. In addition, easier data access can lead to confusion in versioning and even damage versioning consistency. Therefore, to maintain the consistency of product versions, the industrial metaverse needs to do a good job in product data management, especially version management. We use multi-authority attribute-based encryption (MA-ABE) for data access control and deploy blockchain in combination with it to make access records traceable, queryable, and non-tamperable to ensure the security of version data and do well in version management. Based on the limitations of the existing industrial production environment after virtualization, we propose a new concept for the future industrial metaverse, which can realize real-time interaction between users and industrial production in the virtual world and can do a good job in the whole lifecycle of industrial products version management and authority management.

Product lifecycle management (PLM) is an open and interoperable application scheme covering the whole process of product life from birth to death. PLM product information is created, managed, distributed, and used across the enterprise and throughout the product lifecycle, integrating elements such as employees, processes, and information. Ideally, all systems and departments involved in a product would have access to the same database: from design, process, manufacturing, using, and maintenance. Product design includes concept generation and drawing design,

design engineers need to determine the product model according to market demand, user needs, and other factors. The process stage needs to complete the verification and analysis of the product according to the design drawings, give the specific manufacturing scheme, and complete the development of the product prototype and field experiments. This procedure is very important to understand the actual condition of the product and what needs to be improved. The manufacturing stage is when the factory produces a version ready for market according to the prototype manufacturing method provided by the design drawings and technicians. In the using stage, product users officially use the products on the market and give some user experience. Operation and maintenance phase includes the circulation of products in the market, as well as after-sales consultation, maintenance, and other services.

A blockchain is essentially a decentralized database that shares data among distributed nodes in a computer network. It includes new application modes of computer technology such as point-to-point transmission, consensus mechanism, and encryption algorithm. The most famous application of blockchain is its role in cryptocurrencies, such as Bitcoin [13]. That's where he came from. Because blockchain is decentralized, it can facilitate peer-to-peer transactions, so no third-party approval is needed to trade or exchange funds.

A blockchain is a chain-like structure that contains the head node of the chain (the first block) and the tail node (the last block). Each block contains many transactions, and each block contains specific information about that particular block. The primary field of a block is the block header for that block. The block is used to link to the address of the previous block and to provide integrity assurance to the blockchain database. Blocks are linked to each other by block header information, which records the hash value of the previous block (the hash value transformed by the hash function) and the hash value of this block. The hash value of this block is recorded in the next new block, thus completing the information chain of all blocks. In most cases, once a new block is completed and added to the end of the blockchain, the data record of that block can never be changed or deleted. This feature ensures that the database is rigorous and cannot be tampered with. Hyperledger Fabric is a famous open-source consortium blockchain platform that is widely used in the blockchain research [14].

Hyperledger Fabric is an enterprise-class licensed blockchain platform different from free public blockchain, meaning that the identities of all participating members are known and verifiable. A Fabric network consists of multiple channels that are a private subset of communication between two or more concrete network members who can conduct transactions privately and in a confidential manner. This approach provides confidentiality for the deployment of smart contracts. Smart contract is a protocol deployed on blockchain, which is an important technical support for blockchain to support transactions. When pre-set conditions are met, smart contracts run automatically, allowing participants to obtain results directly without the need for redundant personnel. Multiple smart contracts can even be coupled together to automate workflows. Peer nodes are fundamental elements and major components of a Fabric network because they store ledgers and smart contracts that encapsulate processes and information shared in the network for subsequent review and access.

By the way, the specific implementation of blockchain is not presented in this chapter.

8.2 Blockchain-Driven Data Management Example

In this subsection, we will demonstrate our motivation by presenting an example of the full lifecycle of an industrial product. At present, the whole lifecycle of an industrial product refers to the whole evolution process of the product, including the different stages of design, process, manufacturing, using, and maintenance. Different products may go through different lifecycles, but these five stages are the ones most products go through, so let's use them as examples.

In traditional PLM, the various life stages of the physical world are usually digitized to generate product models. A product model is a digital description and representation of product information, such as design drawings, description documents, and so on. In a traditional enterprise information system, the product model is often not unified and scattered in different departments or applications. For example, there are some differences in content, format, and form between the geometric model in product design and the process model in the process stage. In different departments, people in different positions are responsible for different phases of products and managing them. Figure 8.1 shows the five different life stages of the product and the people responsible for them. However, the reality of product design is not plain sailing. If the original version that the designers designed was version 1.0, the version that would eventually be sold in the market would be version 1.x or even version 2.x, version 3.x, and so on. As shown in Fig. 8.1, the plain solid arrows

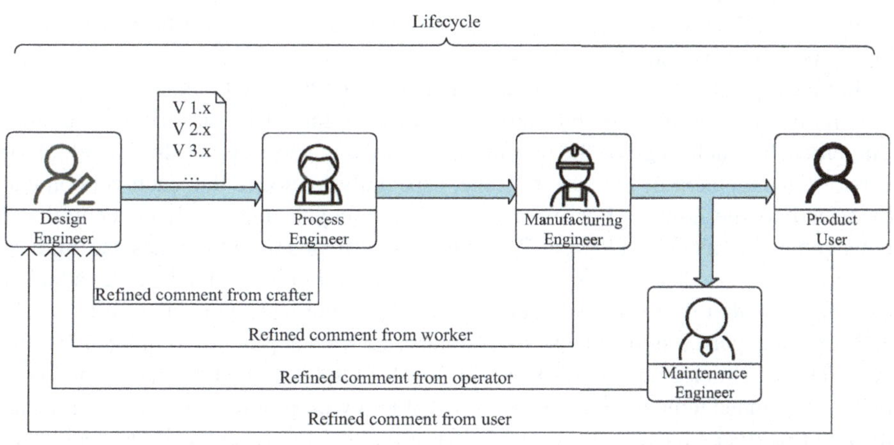

Fig. 8.1 The product lifecycle and version alteration

indicate the version alteration that a product may face during its lifecycle. After the design engineer hands the version data to the process engineer, the process engineer may discover the design mistake and ask the design engineer to redesign it. During manufacturing of version 1.x or 2.x, the manufacturing engineer found that the design drawings or prototyping methods were facing difficulties in actual mass production and went back to the design phase for a redesign. Later, product users were given a stable version of the product, only to find that the actual use of the product did not meet their requirements. The designers redesigned the product after taking into account user feedback. The operation and maintenance engineer will collect product market research and all kinds of maintenance and feedback opinions to design engineers. In the end, a final version was designed and sold in mass production. It is worth noting that although versions 1.x, 2.x, or 3.x eventually took over the market, data from previous versions of the product needs to be retained because they can meet different needs. This is product lifecycle management and version management as we described earlier.

In the industrial metaverse, physical models, sensor data and digital twinning technology can be used to simulate from multiple dimensions and complete mapping in virtual space, to reproduce the whole lifecycle process of corresponding entities in the digital world. In traditional enterprises, it is difficult for specialists to communicate and write in different life stages of products, different departments, and different development stages of products, forming information islands between each other, but in the industrial metaverse is different. As we mentioned in Sect. 8.1, in the industrial metaverse, the physical distance between different life stages will disappear and the real-time interaction between different engineers or users will be greatly enhanced. Therefore, the responsible personnel of different stages can freely access, query, and even modify the product data of other stages. It is important to note that the lifecycle of a product often has interaction requirements only with the phases that are closely related to it. For example, engineers in the design stage will check and even correct the product data in the process stage, and engineers in the maintenance stage will pay more attention to the production data and after-sales service data. Therefore, authority management will be an important problem to be solved in the highly digitized industrial metaverse.

For each phase of product data, data changes, whether due to changes in product requirements or corrections by staff, need to be retained in the original version. Therefore, for each stage of the product lifecycle, it is necessary to do a good job of version management of the current stage, especially to ensure that each version can be traced to the source. For existing product lifecycle management, centralized management is usually adopted. The administrator of the system has the highest authority over the data at each stage. Once the central node is attacked, the attacker can forge an identity for illegal access, resulting in the leakage of files and data. In addition, the permission change of ordinary personnel can be completed by the central node, which leads to the excessive power of the central node and confusion in permission management. When there is a problem with product data, it is likely that it cannot be traced back to the owner. Therefore, how to prevent a single node from having too much access management permission is very important for the life cycle management of industrial products, which is also necessary for the industrial metaverse.

8.3 Attribute-Based Intelligent Data Management Over Industrial Metaverse

In this section, we will give an overview to introduce the framework of the industrial metaverse as well as the data-sharing process.

First is the components involved in the industrial metaverse. As shown in Fig. 8.2, we divide the components in the industrial metaverse including central authority, attribute authority, role, product version database, and blockchain. We describe each of these components in detail below:

- **Central authority (CA)**: The CA plays an important role in the initial phase of industrial metaverse by generating an initial parameter. It is worth noting that CA has nothing to do with the generation of attribute key, as well as the generation of other keys. It is also away from attributes.
- **Attribute authority(AA)**: Attribute authority is the character responsible for issuing attributes in attribute-based encryption. Due to the decentralized design,

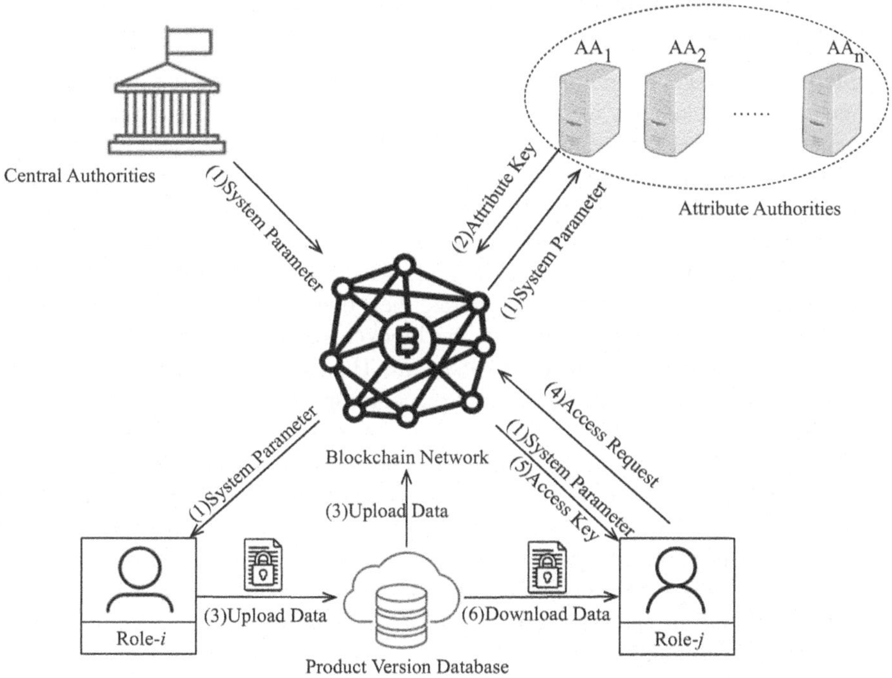

Fig. 8.2 The framework of industrial metaverse

we split the user's attribute set, and different attribute sets are kept by different AAs.

- **Role**: In the industrial metaverse, the role is a broad concept, which refers not only to consumers and users of industrial products but also to managers who are responsible for different stages of the product lifecycle, such as design engineer, manufacturing engineer, and maintenance engineer. In MA-ABE, each user has a unique and distinct global identifier (GID). With the GID, the role can interact with the AA to obtain the decryption credentials for the corresponding attributes.
- **Product version database (PVD)**: As we mentioned in Sect. 8.2, it is very important to do a good job on version management in the industrial metaverse. Therefore, a database dedicated to storing product version data is essential. It is worth noting that the product version data is stored in the PVD in the form of ciphertext. Different industrial metaverse roles access the ciphertext from it.
- **Blockchain**: As Fig. 8.2 shows, the above components are all connected to a blockchain network. As a communications hub, the blockchain interacts with multiple components to exchange data and store access records. At the same time, blockchain stores some public parameters such as public keys and information related to ciphertext in PVD.

Next, we will describe the data-sharing process in our industrial metaverse corresponding to Fig. 8.2. The serial numbers on the arrow correspond to different steps in order. The explanatory note on the arrow is a concise instruction. The two roles in Fig. 8.2 are examples and can be replaced by any other role. The process in detail is as follows:

- **Step 1**: The first is the initialization of the entire system. In the original industrial metaverse, blockchains were deployed and central authority was established. To generate other keys, the central authority randomly generates a security parameter and generates the system parameter from this security parameter. The CA then uploads this system parameter to the blockchain. All other components in industrial metaverse can get this system parameter from the blockchain network.
- **Step 2**: Different attribute authorities and roles have been added to the industrial metaverse. These AA and roles generate their own public and private key pairs based on the system parameter on the blockchain. The AA and the role then upload their public keys to the blockchain. The blockchain then generates the attribute key of a specific attribute according to its corresponding AAs' public keys.
- **Step 3**: A role in an industrial metaverse, such as a design engineer or process engineer, that designs a version of a product and then gets the version data of the product. It then encrypts and uploads the product version data to the PVD based on the system parameter and public keys of AA retrieved from the blockchain. The PVD also uploads the data information to blockchain.
- **Step 4**: When an actor in another industrial metaverse wants access to other product version data, he will send an access request to the blockchain.

- **Step 5**: After monitoring the access quest of role, the blockchain informs AA to provide the access key. According to the role's attribute, AA uses its private key to issue an access key to the role.
- **Step 6**: The role that wants to access the product data gets the access key and the corresponding ciphertext from the blockchain. This role decrypts ciphertext based on the system parameter and attribute token to obtain version data.

8.4 A Blockchain-Driven Method for Product Data's Version Management in Industrial Metaverse

Our blockchain-driven version management approach adds multiple trusted nodes to the blockchain, including attribute institutions and industrial metaverse roles. In the process of version data sharing, role, product version database, and AA interact with the blockchain through smart contracts to obtain the system parameter and exchange tokens and ciphertext. We first introduce the preliminaries and content of smart contracts. Then we give a detailed description of the data exchange in phases. Finally, we give a brief security analysis of the proposed method in the industrial metaverse.

8.4.1 Preliminaries of Version Management in Industrial Metaverse

In the blockchain-driven method for product data's version management, two technologies related with access control are adopted in the data-sharing process. Here is preliminary knowledge of them and their functions in the industrial metaverse.

Multi-authority Attribute-Based Encryption scheme. Attribute-based encryption (ABE) encrypts messages based on attributes without caring about the identity of the receiver. Only users meeting attribute requirements can decrypt the ciphertext, ensuring data confidentiality. In addition, user keys in ABE are related to random polynomials or random numbers, so the keys of different users cannot be combined to prevent users from conspiring to attack. Basic ABE cannot support flexible access control policies. However, the common attribute-based encryption has only a single authorization authority, which is inefficient and unsafe in the decentralized industrial metaverse. Therefore, MA-ABE is proposed. A classic MA-ABE scheme, which is proposed by Lewko A. and Waters B [15], contains four algorithms including Setup, Encrypt, KeyGen, and Decrypt.

1. GlobalSetup(λ) $\rightarrow$ GP. A random security parameter λ is initialized at the beginning. The global setup algorithm takes this parameter to generate global parameters GP for the system. This global parameter is initialized within the

central authority set up in the industrial metaverse and provided to each property agency and various roles in the industrial metaverse for secret key generation.

2. Authority Setup (GP) → SK, PK. Each authority takes GP as input from the system and then gets a secret key and public key pair (SK, PK) through the authority setup algorithm. The key pair is mainly responsible for the encryption and transmission of data in the industrial metaverse. The private key is responsible for data encryption, and the public key is responsible for generating decryption keys for roles in the industrial metaverse.
3. Encrypt(M, (A, ρ), GP,{PK}) → CT. Taking a target message, an access structure (A,ρ), the global parameter GP and the specific attribute public key PK, CT comes out after running Encrypt algorithm. Here, M stands for product data, especially the data of various versions. Due to the open environment in the industrial metauniverse, data needs to be stored and transmitted encrypted.
4. KeyGen(GID, GP, a, SK) → $(K_{a,\ \mathrm{GID}})$. The key generation algorithm takes in an identity GID, the global parameters GP, an attribute a belonging to some authority, and the secret key SK for this authority. It produces a key $(K_{a,\ \mathrm{GID}})$ for this attribute, identity pair. Roles in the industrial metaverse can obtain decryption keys based on their own attributes, such as whether they are design engineers or other members of a product lifecycle.
5. Decrypt(CT, GP, $\{K_{a,\ \mathrm{GID}}\}$) → M. The decryption algorithm converts the ciphertext CT to M by taking the global parameter GP and the unique decryption key related to GID and a specific attribute a. If the attributes cannot meet the access structure contained in the ciphertext, the decryption fails. Otherwise, M will be decrypted successfully. In the industrial metaverse, as we mentioned in 3, only people who meet certain attributes can access attributes owned by other people. If the secret key is incorrect, the ciphertext cannot be decrypted.

Shamir Secret-Sharing scheme. Shamir's secret-sharing scheme is a widely used cryptographic scheme that divides a secret into multiple parts and delivers them across an insecure network. Due to the effectiveness of this scheme, several highly sensitive information like private keys or biometric data can be protected in this way. In general, the secret-sharing scheme will set two values of two different sizes before splitting the secret. The larger value is the total score of the secret segmentation, and the smaller value is called the threshold. As the name implies, the original secret can only be recovered if the number of pieces together reaches a threshold. When the fragments cannot be assembled to meet the threshold, no information about the original secret is disclosed. In our proposed industrial metaverse, we use this method to ensure that the blockchain has enough attributes to compute the attribute key.

A Shamir secret-sharing scheme mainly contains two steps as follows:

1. Share: When given a secret x_0, construct a random polynomial $F(x) = x_0 + a_1x^1 + a_2x^2 + \cdots + a_{t-1}x^{t-1}$. Then, substitute into n different x value to get a set of value pairs $\{(x_1, y_1), (x_2, y_2), \cdots, (x_n, y_n)\}$.
2. Recovery: To recover the secret x_0, at least t of value pairs need to be obtained. After satisfying the threshold t, the secret can be calculated by the LaGrange interpolation following the equation below:

$$x_0 = \sum_{i=1}^{t} \frac{y_i \prod_{1 \le j \le t, j \ne i} - x_j}{\prod_{1 \le j \le t, j \ne i} (x_j - x_i)}$$

8.4.2 Smart Contract of Blockchain for Version Management

To ensure the consistency of version data, this work introduces smart contracts, which are deployed on the blockchain during the initialization phase. The contract is invoked when a role wants to initiate a request for access to the version data.

Contract 8.1 gives the process of assignment to different attributes defined in attribute allocation contract. Because only the CA has the right to initialize the industrial metaverse, the contract will verify the sender first. In the blockchain, the attribute is stored in the form of key–value pairs. For instance, the key is a, with two values that are authoritylist and threshold. The former one indicates the affiliation AA of a and the latter one represents the minimum value to recover the public key of a.

Contract 8.1: Attribute Allocation

procedure AttributeAllocation(a, AAs, t)
if Sender=CA **then**
    $attribute[a].\ authoritylist \leftarrow AA(s)$;
    $attribute[a].\ threshold \leftarrow t$;
    end if
end procedure

Contract 8.2 shows the generation process of the attribute public key of a. As a communication center and trusted platform, blockchain is responsible for storing each public key generated by each AA during system initialization. Since the attribute public key fragment is only available to AA, the sender should be verified as AA at the beginning of the contract. During the initialization of the industrial metaverse, the public keys corresponding to each attribute for encryption are divided into n parts, which need to meet at least t parts to recover. Therefore, after collecting all the public keys of the AA that manages A, the smart contract can meet the conditions and generate the attribute public key PK_a of a.

Contract 8.2: Key Setup

function KeySet(a, AA)
    for AA_i in all AAs **do**
        if $AA_i = AA_a$ **then**
            $count[a] + +$;
            if $count[a] = attribute[a].\ threshold$ **then**
                generate PK_a according to shamir secret sharing;
            end if
        end if

(continued)

end for
  return PK_a
end function

After initialization and attribute public key generation, the blockchain prepares for version data sharing. Contract 8.3 describes the upload of version data. Unlike MA-ABE in Sect. 5.1, this method additionally initializes the public and private key pairs for the role. When role$_i$ signs locally with its private key, it uploads the signed ciphertext to the blockchain along with its public key. The PVD then stores this information. Contract 8.4 is the acquisition of version data. When the receiver is the role$_j$ sending the request, the contract sends the ciphertext and the public key of role$_i$ to it. In addition, the contract calculates the decryption key for role$_j$ based on SK.

Contract 8.3: Upload Version Data

procedure UpData(CT_{sign}, rsk_i)
  if Sender=Role$_i$ **then**
  $dataHash \leftarrow getData(CT_{sign}, rsk_i)$;
    end if
end procedure
procedure SaveData($dataHash$)
    if Receiver=PVD **then**
      $dataHash$ saved;
    end if
end procedure

Contract 8.4: Get Version Data

procedure GetVerData($dataHash$, SK)
  if Recevier=Role$_j$ **then**
    $(CT_{sign}, rpk_i) \leftarrow getCT(dataHash)$;
    $K_{a,\ GID} \leftarrow getKey(SK)$;
  end if
end procedure

8.4.3 The Blockchain-Driven Method for Product Data's Version Management

In this subsection, we will elaborate the whole process of the version management approach from the initialization of the industrial metaverse, version data encryption, token generation, and version data decryption.

8.4.3.1 Initialization

In the initialization phase, the global parameter is generated by CA. Then, it will be uploaded to the blockchain and other components in industrial metaverse generate their public and private keys according to this parameter.

GlobalSetup(λ) $\rightarrow$ GP: CA takes charge of the GlobalSetup algorithm in the initialization. It takes security parameter λ as input. The global parameters GP become outputs. In the global setup, a bilinear group G of order $N = p_1p_2p_3$ is chosen. The global public parameters, GP, are N and a generator g_1 pf G_{p_1}. Moreover, a one-way hash function $H : \{0,1\}^* \rightarrow G$ that maps global identities GID to elements of G is described. Then the CA uploads the global parameters GP to the blockchain.

AASetup(GP) $\rightarrow$ (sk_i, pk_i): After AAs joining the industrial metaverse, they will be allocated a set of different attributes following the Contract 8.1. The AASetup algorithm outputs a pair of key (sk_i, pk_i) according to the global parameters GP for AA_i. For each attribute a managed by the authority, the authority randomly chooses two exponents α_i, $y_i \in \mathbb{Z}_N$ and publishes $\mathrm{pk}_i = \{e(g_1, g_1)^{\alpha_i}, g_1^{y_i} \forall i\}$ as the public key. The secret key $\mathrm{sk}_i = \{\alpha_i, y_i \,\forall\, i\}$ will be kept by AA_i. The public key pk_i will be saved in the blockchain.

RoleSetup(GP) $\rightarrow$ (rsk, rpk): The RoleSetup algorithm takes global parameters GP as input. A pair of keys (rsk, rpk) for a role will become the output later. The rsk is a random value from $\mathbb{Z}_N$ and rpk $= g^{\mathrm{rsk}}$. The public key rpk is also saved in the blockchain.

PVDSetup(GP) $\rightarrow$ (vsk, vpk): The PVDSetup algorithm is just like RoleSetup. It takes global parameters GP as input and then generates a pair of keys (vsk, vpk) for the PVD. These keys are mainly used to encrypt the CT to get the datahash.

KeySet(pk_i) $\rightarrow$ PK_a: The KeySet algorithm is introduced in the Contract 8.2. When the blockchain collects enough public keys pk_i from different AAs, the attributes key PK_a will be calculated by the Shamir secret-sharing scheme.

8.4.3.2 Version Data Encryption

Encrypt($M, (A, \rho), \mathrm{GP}, \mathrm{PK}_a$) $\rightarrow$ CT: The role that upload the product version performs *Encrypt* algorithm. The encryption algorithm takes in the version data M an $n \times l$ access matrix A with ρ mapping its rows to attributes. It chooses a random $p \in \mathbb{Z}_N$ and a random vector $v \in \mathbb{Z}_N^l$ with p as its first entry. Let λ_x denote $A_x \cdot v$, where A_x is row x of A. It also chooses a random vector $w \in \mathbb{Z}_N^l$ with 0 as its first entry. We let ω_x denote $A_x \cdot w$. For each row A_x of A, it chooses a random $r_x \in \mathbb{Z}_N$. The ciphertext is computed as:

$$\begin{aligned} C_0 &= Me(g_1, g_1)^s \\ C_{1,x} &= e(g_1, g_1)^{\lambda_x e(g_1, g_1)^{\alpha_{\rho(x)} r_x}} \\ C_{2,x} &= g_1^{r_x} \\ C_{3,x} &= g_1^{y_{\rho(x)} r_x} g_1^{\omega_x} \forall x \end{aligned}$$

Sign(CT) → $\mathrm{CT}_{\mathrm{sign}}$: Role_i signs the ciphertext with its secret key rsk_i. Then the signed ciphertext $\mathrm{CT}_{\mathrm{sign}}$ will be uploaded to the blockchain by a simple smart contract.

8.4.3.3 Token Generation

When a role launch a request to access the product version data, the blockchain will inform AAs assuming related attributes to generate the decryption key of attribute a in blockchain. The blockchain will calculate the decryption key for the role and get ready to send it.

KeyGen(GID, a, SK, GP) → $K_{a,\ \mathrm{GID}}$. To generate an access key for the role with GID on attribute a belonging to its attribute key, the authority computes:

$$K_{a,\mathrm{GID}} = g_1^{\alpha_a} H(\mathrm{GID})^{y_a}$$

8.4.3.4 Version Data Decryption

VerifySig($\mathrm{CT}_{\mathrm{sign}}$) → CT: Before the decryption algorithm, role_j utilizes the public key rpk_i of role $_i$ to verify the authenticity of CT.

Decrypt(CT, $K_{a,\ \mathrm{GID}}$, GP) → M: The decryption algorithm takes ciphertext CT, decryption token $K_{a,\ \mathrm{GID}}$, and global parameters GP as input. It outputs version data M. The ciphertext is assumed to be encrypted under the access matrix an access matrix (A, ρ). To decrypt, role $_j$ will get H(GID) from the random oracle. If role $_j$ has the secret keys $K_{\rho(x),\ \mathrm{GID}}$ for the subset of rows A_x of A such that (1, 0,..., 0) is in the span of these rows, then the decryptor of role_j proceeds as follows. For each such x, the decryptor computes:

$$\frac{C_{1,x} e(H(\mathrm{GID}), C_{3,x})}{e\left(K_{\rho(x),\mathrm{GID}}, C_{2,x}\right)} = e(g_1, g_1)^{\lambda_x} e(H(\mathrm{GID}), g_1)^{\omega_x}$$

The decryptor then chooses constants $c_x \in \mathbb{Z}_N$ such that $\sum_x c_x A_x = (1, 0, \cdots, 0)$ and computes:

$$\prod_x \left(e(g_1, g_1)^{\lambda_x} e(H(\mathrm{GID}), g_1)^{\omega_x} \right)^{c_x} = e(g_1, g_1)^s.$$

Then the version data M can be calculated as:

$$M = \frac{C_0}{e(g_1, g_1)^s}$$

8.5 Evaluation and Security Analysis

To test the performance of the proposed blockchain-driven method for product data's version management in industrial metaverse, an experimental environment of a consortium blockchain is set up in this section to conduct an experimental evaluation of the method. Table 8.1 shows the experimental environment of this chapter. Our experimental data comes from the design data of excavators and other engineering instruments of Xuzhou Construction Machinery Group (XCMG), China's largest construction machinery manufacturer. The data set contains the overall design drawing data of engineering instruments and the data of parts.

In the blockchain-driven method, data uploading and smart contract deployment belong to the write operation on blockchain, while logs and data viewing belong to the read operation on blockchain. The read/write speed on blockchain is not only related to the hardware environment but also to the number of nodes in the blockchain network. This chapter configures a sorting node with 4,6,8,10 peer nodes to test the speed of reading and writing blockchain under different conditions. The tests tested the speed of reading and writing blocks in different environments and the time it took to complete the transaction by simulating the continuous sending of access requests to the blockchain at different speeds. The result is shown in Fig. 8.3a and b from the above experimental analysis, it can be seen that the efficiency of the traceable access control mechanism based on blockchain is closely related to the number of nodes in the alliance chain. The number of nodes affects the scale of the access control system. Select an appropriate number of nodes on the premise of ensuring the scale and efficiency of access control. In addition, the length and structure of the request also affect the efficiency of the read-and-write blockchain. To improve the efficiency of the access control system, the data structure of each request should be optimized as much as possible to reduce the transaction delay.

Table 8.1 Experimental environment

Environment	Description
Software	Hyperledger fabric 2.4.5 and Go 1.18.4
Hardware	MacOS Catalina 10.15.7, Intel Core i7(4-core,2.3GHz)

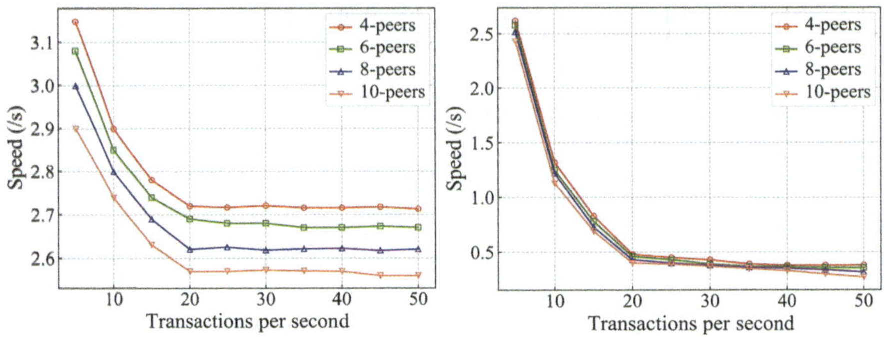

(a) (b)

Fig. 8.3 Read and write speed on blockchain under different number of peers. (**a**) Read speed. (**b**) Write speed

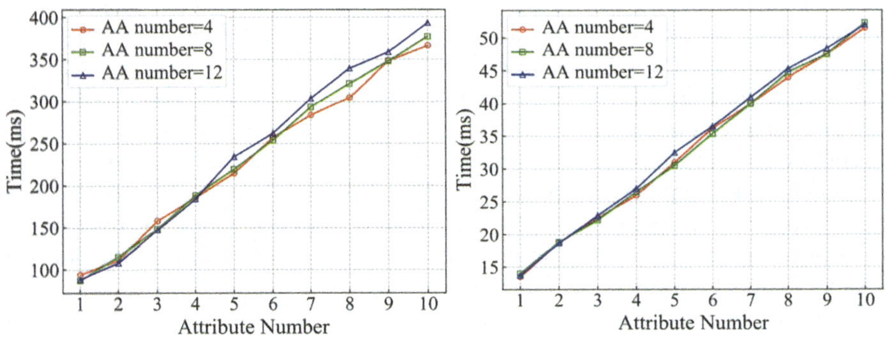

(a) (b)

Fig. 8.4 Read and write speed on blockchain under different number of peers. (**a**) AASetup time. (**b**) KeyGen time

Figure 8.4a and b, respectively, show the relationship between the running time of AASetup algorithm and KeyGen algorithm and the number of management attributes when the number of authorities is 4, 8, and 12. This group of experiments has been run for 50 times, and the values in the figures are the average values of the experimental results. As can be seen from the Fig. 8.4, the running time of AASetup algorithm and KeyGen algorithm has a linear relationship with the attributes of management, and the running time of the algorithm increases linearly with the more attributes of management. The number of authorities has little effect on the running time of the algorithm. As the number of authorities increases, the running time of AASetup algorithm increases slightly, and the increase is more obvious when the number of attributes is large. However, the relationship between the running time of KeyGen's algorithm and the number of authorities is weaker, and this is true regardless of the number of attributes.

Fig. 8.5 The XCMG Global Service System that the version management method applies to

Here, it is worth mentioning that the prototype of the industrial metaverse used in this chapter is originated from a real industrial platform of XCMG Global Service System as shown in Fig. 8.5 (http://xgss.xcmg.com/#/login?redirect=%2F). The blockchain-driven method for product data's version management has been introduced into this platform for enhancing the consistency of its PDM system. The blockchain-driven method presented in this chapter is the first step toward a real operation of product data's version management in an industrial metaverse. More and more blockchain-driven operations would be developed in our future research work with a long-term goal of setting up a well-organized ecological system for supporting various industrial metaverse applications.

Some of the details of the previous system architecture will be added to demonstrate the basic security of our system. The superficial analysis is as follows:

1. **Traceable Source**: When one role accesses the different phases of product data stored in the product version database by another role via blockchain transactions, the blockchain logs the access on the chain for verification. The role responsible for the product data in this phase can view the visitor authorization and access process via blockchain and verify that the server is open to the role in accordance with the access control policy. Cloud servers can also use blockchain to prove that data access processes comply with access control policies.
2. **Unforgeability**: Due to the immutable nature of the blockchain, access logs recorded on the blockchain cannot be tampered with or even forged by any role or server. Immutability and unforgeability effectively eliminate security problems caused by a server being attacked or bribed, because access logs expose illegal rejections by dishonest servers.
3. **Data Privacy**: The Hyperledger Fabric blockchain platform supports transaction privacy, where roles obtain tokens through attribute authorization to access data.

Although the cloud server stores ciphertext, it cannot decrypt the ciphertext without the role private key, thus ensuring the privacy of data storage.

8.6 Summary

In the whole lifecycle of a product from generation to retirement, various autonomous roles such as design engineers, process engineers, manufacturing engineers, maintenance engineers, product users, etc., need to exchange data frequently to promote the iteration and update of the product, resulting in the requirements of version management. We combine the concept of metaverse with industry and propose the concept of industrial metaverse, which maps various roles to virtual worlds to facilitate the sharing of product version data. Due to the open characteristic of industrial metaverse, the version management of product data is more urgent and strict. We propose a blockchain-driven method for product data's version management that utilizes multi-authority attribute-based encryption for security and uses blockchain as a channel for data sharing. We deployed our blockchain experimental environment using Hyperledger Fabric, conducted experiments based on real industrial data, and tested the performance of our approach from different perspectives.

References

1. Liu DT, William X, Xu. (2001) A review of web-based product data management systems. Comput Ind 44(3):251–262
2. Panetto H, Dassisti M, Tursi A (2012) ONTO-PDM: product-driven ONTOlogy for product data management interoperability within manufacturing process environment. Adv Eng Inform 26(2):334–348
3. Zhao X, Liu C (2007) Version management in the business process change context. In: International Conference on Business Process Management, Springer, pp 198–213.
4. Yang Q, Zhao Y, Huang H, Xiong Z, Kang J, Zheng Z (2022) Fusing blockchain and AI with metaverse: a survey. IEEE Open J Comput Soc 3:122–136
5. Mozumder MAI, Sheeraz MM, Athar A, Aich S, Kim H-C (2022) Overview: technology roadmap of the future trend of metaverse based on IoT, blockchain, AI technique, and medical domain metaverse activity. In: 2022 24th International Conference on Advanced Communication Technology (ICACT), pp 256–61
6. Ifdil I, Situmorang DDB, Firman F, Zola N, Rangka IB, Fadli RP (2023) Virtual reality in metaverse for future mental health-helping profession: an alternative solution to the mental health challenges of the COVID-19 pandemic. J Public Health 45(1):e142–e143
7. Tang F, Chen X, Zhao M, Kato N (2022) The roadmap of communication and networking in 6G for the metaverse. IEEE Wireless Communications, (Early Access). https://doi.org/10.1109/MWC.019.2100721
8. Kechadi T, Chen L, Aung N, Ning H, Atzori L, et al (2022) Edge-enabled metaverse: the convergence of metaverse and mobile edge computing. arXiv:2205.02764
9. Duan H, Li J, Fan S, Lin Z, Wu X, Cai W (2021) Metaverse for social good: a university campus prototype. In: Proceedings of the 29th ACM International Conference on Multimedia, pp 153–61

10. Rosen R, Von Wichert G, Lo G, Bettenhausen KD (2015) About the importance of autonomy and digital twins for the future of manufacturing. Ifac-Papersonline 48(3):567–572
11. Li X, Liu H, Wang W, Zheng Y, Lv H, Lv Z (2022) Big data analysis of the internet of things in the digital twins of Smart City based on deep learning. Futur Gener Comput Syst 128:167–177
12. Kraska T, Stonebraker M, Brodie ML, Servan-Schreiber S, Weitzner DJ (2019) SchengenDB: A data protection database proposal. In: Heterogeneous Data Management, Polystores, and Analytics for Healthcare - VLDB 2019 Workshops, Poly and DMAH, Los Angeles, CA, USA, August 30, 2019, Revised Selected Papers, pp 24–38
13. Nakamoto S (2008) Bitcoin: a peer-to-peer electronic cash system. Decentralized Business Review
14. Androulaki E, Barger A, Bortnikov V, Cachin C, Christidis K, De Caro A, Enyeart D, et al (2018) Hyperledger fabric: a distributed operating system for permissioned blockchains. In: Proceedings of the Thirteenth EuroSys Conference, pp 1–15
15. Lewko A, Waters B (2011) Decentralizing attribute-based encryption. In: Annual International Conference on the Theory and Applications of Cryptographic Techniques, pp 568–588

Chapter 9
Mobile Business Task Management Over Industrial Internet

Abstract In the past few years, latency-sensitive task computing over the Industrial Internet has played a key role in an increasing number of intelligent applications, such as intelligent self-driving vehicles and unmanned aircraft systems. The edge computing paradigm provides a basic functional infrastructure for across-domain business task computing on distributed edge servers. With this observation, a trade-off between the mobile devices and the fixed-edge servers is needed to run moving task computing in a low-latency way. In view of this challenge, this chapter investigates the framework of mobile computing and Industrial Internet applications in managing business tasks. The framework considers the location and mobility of workers in industrial settings, enabling more efficient task management. The server selection method is then introduced. This method concentrates on the novel location-aware environment. The majority of users can receive edge service with the lowest possible system cost.

In the past few years, latency-sensitive task computing over the Industrial Internet has played a key role in an increasing number of intelligent applications, such as intelligent self-driving vehicles and unmanned aircraft systems. The edge computing paradigm provides a basic functional infrastructure for across-domain business task computing on distributed edge servers. With this observation, a trade-off between the mobile devices and the fixed-edge servers is needed to run moving task computing in a low-latency way. In view of this challenge, this chapter investigates the framework of mobile computing and Industrial Internet applications in managing business tasks. The framework considers the location and mobility of workers in industrial settings, enabling more efficient task management. The server selection method is then introduced. This method concentrates on the novel location-aware environment. The majority of users can receive edge service with the lowest possible system cost.

W. Dou et al., *Intelligent Industrial Internet Systems*,
https://doi.org/10.1007/978-981-99-5732-3_9

9.1 Location-Aware Business Task Computing Over Industrial Internet

With the rapid development of mobile applications, terminal device capacity is frequently insufficient to support distributed complex business task computing, particularly for some domain-across task collaborations. To solve this problem, computing duties are routinely offloaded to a cloud or data center. By utilizing cloud computing capability, terminal devices may manage complicated activities for an expanding variety of intelligent apps [1]. During the operation of some IIoT applications, such as intelligent self-driving cars and unmanned aircraft systems, several computational operations are carried out. Task computing typically has stringent latency constraints in reality. However, the network frequently experiences extremely high levels of data traffic as a result of the coupled sensors or terminal devices utilized in these systems. For these latency-sensitive IIoT applications, this typically has negative effects. To meet this problem, the edge computing paradigm provides a fundamental functional framework for distributed edge servers running cross-domain business task computing [2, 3]. To operate complicated business task computing with low latency, however, a trade-off between the mobile device, the edge servers, and the cloud data center is required due to an edge server's limited processing and storage capabilities.

Edge computing is a cutting-edge paradigm that offers new opportunities for mobile device server selection while also posing new problems [4, 5]. The location of edge servers is fixed, and their service area is limited, so tasks will transfer between edge servers as vehicles move [6, 7]. The effectiveness of the system is significantly impacted by different edge server selection techniques. Real-time response is necessary for unmanned aircraft systems as well. On edge servers, processing administrative chores for unmanned aircraft systems can guarantee prompt network assistance. The edge servers' capacity is constrained though. In the event of additional business tasks, incorrect server selection will lead to edge server overload and lower service quality [8, 9]. A number of industries have recently employed server selection. The network field has also intensively investigated server selection to profit from its advantages in lowering network latency, load balancing, and other benefits.

Despite the fact that the server selection problem has drawn a lot of attention [10, 11], the current research does not adequately address four major issues. First, as a result of the devices' mobility and connection with servers and other devices, traditional edge computing has become a location-aware environment, and the storage consumption during communication cannot be ignored [12]. Second, the second assumption made by service providers is that customers would constantly connect to the best edge server along their route. However, a cost-effective task transfer approach should be considered due to the constrained capacity of edge servers and the possibility of job transfer. Third, the majority of studies focus on methods for choosing servers for users who are working alone or who are executing a single job [13–16]. Few studies consider the views of service providers regarding the stress on edge servers. Some edge servers may get overloaded as a result of the unorganized rivalry between users for resources. Fourth, a centralized approach or a set protocol cannot be used to tackle the multiuser server selection problem. Making a distributed service transfer algorithm is the solution.

9.2 Edge Server Selection in Mobile Business Task Management

An illustration of a location-aware environment is shown in Fig. 9.1. Most of the components in this scenario are mobile devices, base stations, and edge servers. Through the base station, services from the edge server are transmitted to the mobile device. Instead of what is generally referred to as an edge computing environment, a location-aware environment allows the base station to determine the user's location. A user device's location varies when it moves, and in a location-aware environment, the base station can pinpoint the user's location even when they are outside the range of the base station.

The edge server performing the user's work is informed of the base station's additional user detections, including the user's direction of travel and data

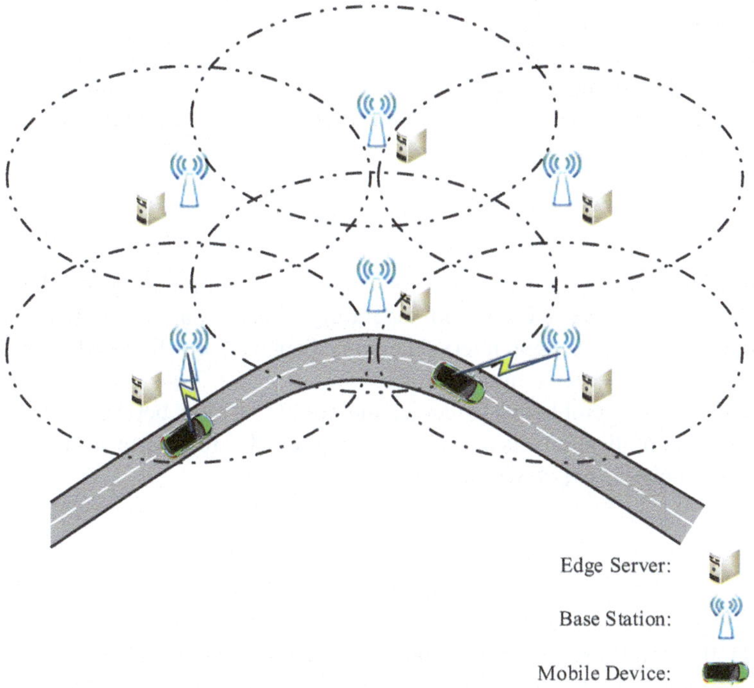

Fig. 9.1 A scenario of a location-aware environment

transmission rate. Based on the perceived information, the edge server that serves the user task can determine whether it needs to be migrated when the user task's location changes and which candidate server is the optimal location to migrate to once the migration is triggered. For instance, Fig. 9.1 shows how a mobile device shifts between edge node service regions while the server selection varies.

The location-aware environment considered in this chapter has a base station layout that is similar to a dense cellular network, as seen in Fig. 9.1. In the subsequent modeling process, BS represents the base station. The user-level server selection problem is also taken into consideration in this study, which calls for the base station to be able to recognize the user's movement direction, speed, and data transmission rate.

s_i is used to represent the edge server and u_j is used to represent the user in the location-aware environment. In addition, the user's tasks are shown as t_j. Only the task's storage resource needs are taken into account in this chapter. The symbols sc_i represent the remaining storage resources. The user task's required storage resources are indicated by the symbol ts_j. Finally, the time is discretized using a time slot model. A time slot is comparable to the idea of a CPU time slice. Time slots are represented by p in the model. The server selection policy between time periods 0 and q is represented by the set α.

Delay limitations come in two different varieties. One is transmission delay, which is the length of time it takes to transmit and receive data during a process. The second is the processing delay, which shows how long it takes a task to complete after it reaches its destination.

Assume that task t_j has a transmission delay of Td_p in time slot p and the interval between u_j and its running server is measured by network hops. In the model, when user task t_j is running on edge server s_i in time slot p, the communication interval count is shown as follows:

$$I\left(t_j, s(p)\right)$$

Assume that a task t_j has a transmission delay of Td_p in time slot p and the interval between u_j and the server it is currently operating on is measured in network hops. In the model, the communication interval count is displayed as follows when user job t_j is running on edge server s_i in time slot p:

In addition, μ is employed to depict the weighted ratio between transmission delay and communication distance. Consequently, the following is how a task's transmission delay is expressed in a time slot p:

$$\mathrm{Td}_p = I\left(t_j, s(p)\right)$$
$$s.t. \quad p \in [0, q]$$

Under the condition of guaranteeing the communication delay constraint, the communication delay t_j in the model shall not exceed the delay threshold. Use th_j to represent the delay threshold constraint and Pd_j to represent the task t_j processing latency to express this relationship. This allows for the following constraint relationships to be discovered.

$$\mathrm{Td}_p + \mathrm{Pd}_j \leq \mathrm{th}_j$$
$$s.t. \quad p \in [0, q]$$

Assume that there are two main costs: one is the communication cost that causes the task to communicate with the server continuously, and the other is the transfer cost when the task is transferring. The communication cost is denoted as com _ cost, and the transfer cost is denoted as tra _ cost. Here, com _ cost is defined as follows.

Assume there are two main costs: the communication cost, which keeps the job in constant communication with the server, and the transfer cost, which occurs when the task is moving. The transfer cost is represented as tra _ cost and the communication cost is denoted as com _ cost. com _ cost is described as follows:

Definition 1 (com_cost) *Suppose there is a task active on the edge network. com _ cost is the storage capacity occupied as a result of the server and mobile device's continuous connection. com _ cost is the result of the communication distance, data transmission rate, and duration.*

According to the definition, the data transmission rate is represented by r, which represents the amount of data transmitted per second. μ represents the length of each time slot in seconds. For a certain task t_j, its com _ cost between time slots 0 and q is notated as com _ cost$_j$. Thus, it can obtain the following.

According to the definition, the amount of data delivered each second is denoted by r, which is also known as the data transmission rate. μ denotes the time in seconds that each time slot lasts. com _ cost$_j$ is noted as being com _ cost between time slots 0 and q for a certain task t_j. As a result, it can get the following:

$$\mathrm{com}_{\cos \mathrm{t_j}} = \mu \cdot r \cdot \sum_{p=0}^{q} I\left(t_j, s(p)\right)$$

Assume that tra _ t represents the price of a single job transfer. The cost is what it would cost to transfer the job from the server it is currently running on to another server. In reality, the task's data size is what mostly determines the transfer cost. However, the user experience will unavoidably be affected to some extent when the job is moved, therefore the data size of the task cannot just be used to calculate the transfer cost. Introduce a weighting factor τ, which reflects the degree of the tolerance of various jobs to the transfer downtime, to capture the influence on user experience when transferring. Thus, for a task t_j with the task size ts_j:

$$\mathrm{tra\_}t_j = \tau \cdot ts_j$$

Definition 2 (*tra _ cost*) *tra _ cost is the storage space occupied by the transfer when a job is moved from the server on which it is executing to another server, given that it is running on the edge network. tra _ cost can be determined by adding up the costs of all transfers and subtracting the job execution time.*

$s(p, p + 1)$ is used to indicate whether user task t_j has been transferred from time slot p to time slot $p + 1$ and the transferring cost of a certain task t_j between time slots 0 and q as tra _ cos t_j.

$s(p, p + 1)$ is used to show whether a user job t_j has been moved between time slots p and $p + 1$ to calculate the cost of moving a specific task t_j between time slots 0 and q.

$$\text{tra\_cos}\,\mathrm{t}_j = \tau \cdot \text{tra\_}t_j \cdot \sum_{p=0}^{q} s(p, p+1)$$

$$\begin{cases} s(p, p+1) = 0, & s(p) = s(p+1) \\ s(p, p+1) = 1, & s(p) \neq s(p+1) \end{cases}$$

Definition 3 (*total _ cost*) *Given a collection of edge network-based tasks. cost is the overall storage space used during its operation. cost is the total of the task's transfer and communication costs during its entire life cycle.*

$$\text{cost} = \sum_{j=1}^{m} \text{sign}(t_j) \cdot (\text{com\_cos}\,\mathrm{t_j} + \text{tra\_cos}\,\mathrm{t}_j)$$

The equation above shows how a series of tasks will be distributed among time slots 0 and q. The problem formulation takes availability into account in addition to cost. The system's availability is noted by *Avi*, and its availability is defined as follows:

Definition 4 (*Avi*) *Avi is the proportion of tasks completed by edge servers to total jobs completed in a second. sign(t_j) serves a symbolic purpose. This means that if a job t_j is running on the edge server at a specific moment sign(t_j) = 1, otherwise sign (t_j) = 0.*

Therefore, the Avi of a particular second for a given set of jobs $T = \{t_1, t_2, \ldots, t_m\}$ is as follows:

$$\text{Avi} = \frac{1}{m} \cdot \sum_{j=1}^{m} \text{sign}(t_j)$$

The optimization of the server selection issue has two basic goals. Maximizing the system's availability is the initial goal. The second is to keep the system's overall cost as low as possible while maintaining availability. The server selection problem's optimization goals are as follows:

$$
\begin{array}{ll}
 & \min \quad \text{cost} \\
s.t. & \max \quad \text{Avi} \\
 & I\left(t_j, s(\text{p})\right) \leq \dfrac{\text{th}_j - \text{Pd}_j}{\mu} \quad ,j \in [1, m]
\end{array}
$$

9.3 Intelligent Edge Server Selection in Mobile Industrial Internet Environment

Calculating the formulas under realistic conditions is difficult because edge computing scenarios are very fluid and many of the formulated parameters depend on future user input. The load on the edge server is not taken into account when the cost model is estimated under ideal circumstances. Direct application of the optimal transfer approach under ideal circumstances might lead to server overload in busy environments. The candidate server priority attained under ideal circumstances is still useful, though. It can serve as a blueprint for creating a practical service transfer method. By utilizing specific scheduling strategies, it is able to address the server selection issue in a multiuser environment when designing the server selection method. Assume that there are no restrictions on the edge server's resources. Since there is no historical data when a task is performed in the cloud, it is possible to move tasks to edge servers. The historical run data of the user's task are known when it is running on an edge server, and this information can be used to decide when to migrate the task and what servers should take priority.

Only single-user tasks with unlimited edge server resources are covered by the aforementioned solution to the server selection issue. In addition, it must be taken into account the server selection issue for multiple user tasks on edge servers when resources are limited. The primary cause of edge server congestion caused by the outdated method is overly competitive users. Under the mechanism, the user task can only send the pretransfer request; once the applicant server has given the user permission, the user can then transfer the job. It is important to note that in the SeeMe method, the edge server consistently sends the user's prerequest. Figure 9.2 displays a straightforward process illustration of the Intelligent edge server selection method: SeeMe.

The algorithm is primarily made up of five pieces that work together. S_{source} represents the user's source edge server, whereas $S_{\text{candidate}}$ represents the candidate edge server to which u_j might transfer. t_j executes the following command:

Step 1: S_{source} identifies the location information of t_j.

Step 2: S_{source} collects all tasks' location information. According to the task's location information, determine whether the task needs to be transferred. If the task reaches the optimal transfer location, S_{source} calculates the priority of the task's candidate servers. Then, S_{source} sends the request to the highest priority

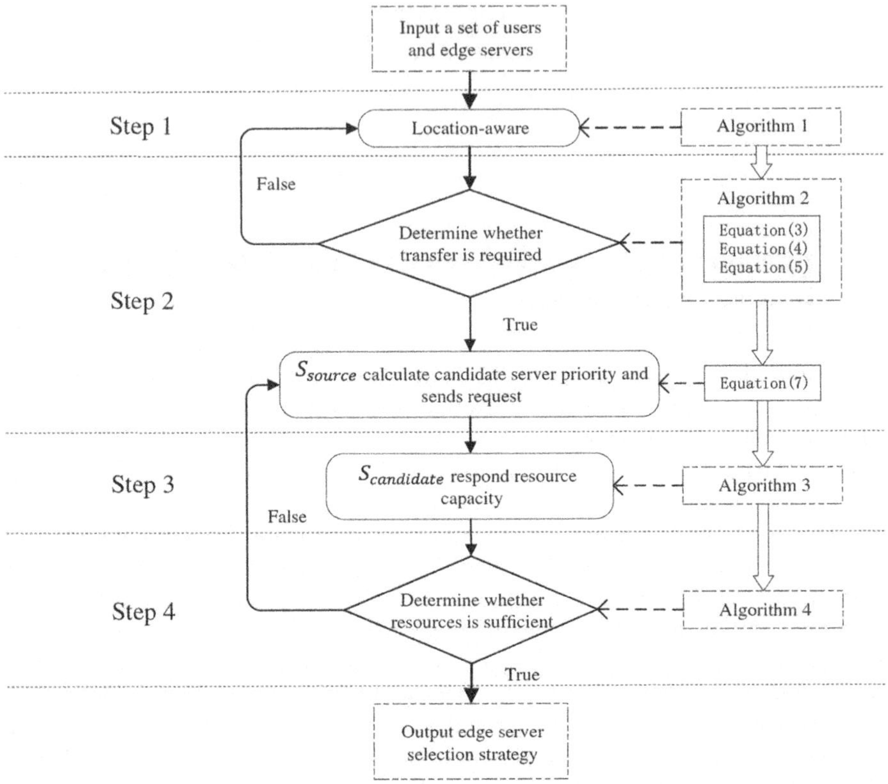

Fig. 9.2 Intelligent edge server selection method: SeeMe

server in the candidate server list. In a time slot, S_{source} will only send the request once for a task. If the task is not in the optimal transfer position, S_{source} will continue identifying the task location information.

Step 3: $S_{\text{candidate}}$ collects the requests in a time slot. Based on the amount of the available resources to $S_{\text{candidate}}$, $S_{\text{candidate}}$ decides the amount of resources to allocate to S_{source} and returns the preallocated amount of resources to S_{source}. For a limited time, $S_{\text{candidate}}$ locks resources.

Step 4: If the preallocated resource is sufficient, S_{source} will decide which tasks to transfer according to the task priority. Otherwise, the tasks that fail to transfer will remove its highest priority server in its candidate server list. Then, these tasks will participate in the transfer in the next time slot in Step 2. The remaining resources of the preallocated resource will be unlocked at the end of the time slot.

9.3.1 Step 1: The Location-Aware

Assume that the user can determine its own movement direction, position, speed, and distance from the source server as well as determine if the base station to which it is connected has moved. The source server receives the aforementioned four bits of information if the user's BS connection changed. As mentioned above, in time slot p, the communication interval of the task t_j is $I(t_j, s(p))$. In p, t_j is running on S_{source}, so $I(t_j, S_{\text{source}}) = I(t_j, s(p))$.

Algorithm 9.1: SeeMe Location-Aware Process

```
1: initialize: BS,t_j,Set time slot is 0
2: for time slot p is in [0,q] do
3:    Check I(t_j, S_source)
4:    if I(t_j, S_source) has changed then
5:       respond I(t_j, S_source), location_j , direction_j , speed_j to S_source
6:    end if
7: end for
```

9.3.2 Step 2: S_{source} Sends Pre-Request

Although the problem is a complex multi-objective optimization problem, the optimization goal of the optimal strategy can be described in the model. Let's relax the requirement that edge servers have a finite amount of resources to make this issue simpler. The global optimal policy in this situation is the one that is best for every user. When transferring data, a number of policies are created to make sure that user tasks' resource demands don't go over the server's maximum capacity. When the edge server resources are limited, the flow depicted in Fig. 9.2 accomplishes the objective of resource allocation for multi-user tasks in an orderly manner. However, there are two additional issues that must be resolved when using this flow. The timing of migration for edge-side tasks is a challenge. The priority of each candidate server on the list of candidate servers for edge-side tasks is the other issue.

Determine the best time for edge-side tasks to initiate migration when resources are not limited in the first place. In this study, the goal of optimization was to reduce the overall system overhead. As a result, the average overhead for a task during each time slot is reduced. As a result, when the task starts to migrate, its location must be at the location with the lowest average overhead. If the task's most recent migration concluded at time slot a and the present time slot is b, then Eq. (9.4) can be used to determine the task's communication overhead at this time, which will be denoted as $\text{com\_cos}\,\text{t}_b^a$. If migration is initiated at this point, the sum of $\text{com\_cos}\,\text{t}_b^a$ and tra_t, where tra_t denotes the overhead of a single migration, can be used to compute the overall overhead in the time slot interval $[a, b]$.

The task has the lowest average cost for each time slot, indicating that is assumed to have the lowest value $\frac{\cos t_b^a}{b-a}$. It is not a convex function, nevertheless, because the user task's mobility and data transfer rate fluctuate arbitrarily. In this case, change the task's historical average for the time slot interval to replace the movement speed and data transfer rate, turning A into a convex function. A simple strategy is to trigger a migration if $\frac{\cos t_b^a}{b-a} \leq \frac{\cos t_{b+1}^a}{b-a+1}$ holds.

The candidate servers' priorities must also be established when a transfer is found to be necessary. Theoretically, any edge server that satisfies the user device communication latency limitation qualifies as a candidate server. To streamline the calculation, just the edge servers on the user's mobile trajectory are considered for choosing the candidate servers. The mobility trajectory of the user device will be covered by the base stations linked to these edge servers.

Calculate the cost of communication to travel from the current location to the candidate server's location for a specific candidate server e_i. The communication cost between two neighboring migrations of the work is estimated to be twice of this value. When the migration task is chosen e_i at this point, the average cost inside each time slot can be assessed. According to their average cost, the prospective servers are prioritized in decreasing order.

Algorithm 9.2: SeeMe in $S_{\mathbf{source}}$

1: initialize: *BS,t,s, timeslot*
2: for time slot p in $[0,q]$ do
3: for t_j running on S_{source} do
4: get $I(t_j, S_{source})$, $location_j$, $direction_j$, $speed_j$ from t_j
5: Calculate r
6: if $\frac{cos\, t_b^a}{b-a} \leq \frac{cos\, t_{b+1}^a}{b-a+1}$ then
7: add u_i to transfer set(Tra_set)
8: end if
9: end for
10: for t_j in Tra_set do
11: calculate the priority of the candidate edge server.
12: end for
13: for i in $[0,n]$ do
14: Calculate how much storage resource the task will demand in total from s_i, then set the value as res_i.
15: end for
16: for i in $[0,n]$ do
17: S_{source} sends res_i to s_i.
18: end for
19: end for

The edge server has two main duties when serving as the source server. It ought to keep track of how many users arrive at the transfer point during each time period, then produce a list of potential servers with a priority for each user based on their distance from the transfer point and other variables like their driving directions. Second, resource pre-requests must be received by each potential server. The total number of user tasks that need to be moved is displayed in lines 2 through 9 of

Algorithm 9.2. The values of r must be calculated here. r is substituted with historical averages in the experiment. Lines 10–12 compute the list of potential servers for each user. The method previously mentioned can be used to determine the order of importance for each candidate server. The total number of resource requests made to each candidate server by the source server are totaled on lines 13 through 15. On lines 16–18, the complete resource requests are sent to all of the candidate servers that are open for selection.

Each user task that needs to be transferred will make an attempt to do so in the decreasing order of candidate server priority during the transfer attempt. If the transfer is unsuccessful, it will attempt to switch to a less important edge server until it is successful. When a pre-request is made, the actual transfer of the user task does not take place. The transfer is carried out after the user task has received the assigned resources, and the transfer does not result in candidate server overloads.

9.3.3 Step 3: $S_{candidate}$ Handles Pre-Requests

Algorithm 9.3: SeeMe in $S_{candidate}$

1: initialize: BS,t,s, set time slot is 0
2: for time slot p in $[0,q]$ do
3:   $sum\_ss = 0$
4:   for i in $[0,n]$ do
5:     $sum\_ss = sum\_ss + res_i$
6:   end for
7:   if $sum\ ss \leq$ Remaining resource of $S_{candidate}$ then
8:     for i in $[0,n]$ do
9:       Allocate res_i storage resources to s_i.
10:     end for
11:   else
12:     for i in $[0,n]$ do
13:       Allocate $\frac{res_i \cdot ss_c}{sum\_ss}$ storage resources to s_i
14:     end for
15:   end if
16:   for i in $[0,n]$ do
17:     Sends the number of allocated resources to s_i.
18:   end for
19: end for

The edge server has two main duties when serving as a candidate server. The primary responsibility is to assess the pre-request data from the statistical source server for each time slot and distribute resources in accordance with the statistics. Algorithm 9.3 counts the total number of resource requests on lines 3–7. When there are sufficient remaining resources, lines 9–12 distribute resources based on the amount applied to each S_{source}; when there are insufficient remaining resources, lines 14–17 distribute resources based on the proportion of applied resources. The second responsibility of the candidate server is to send the allocated amount of resources to the edge server of each requested resource. In lines 19–21, $S_{candidate}$ sends the allocated number of resources to the related source server.

9.3.4 *Step 4:* S_s *Performs the Actual Transfer*

The order of the transfer steps can be determined by the importance of the user tasks being transferred. The priority of user tasks can all be determined using the change in global cost following transfer.

Algorithm 9.4: SeeMe in S_{source} Do the Actual Transfer

```
initialize: BS,t,s, set time slot is 0
for time slot p in [0,q] do
    for t_j in Tra _ set do
        Calculate Priority for t_j
    end for
    for t_j in Tra _ set do
        Do actual transfer in order of user priority.
    end for
end for
```

The proposed method's temporal complexity is calculated as O(nm). The scale of the issue is n and there are m edge servers. User information perception and computation together account for the majority of the algorithm's operational time complexity. Sensing the location data within each time slot has an O(1) time complexity. During the task lifetime, multiple base stations are passed, making the time complexity of this part O(m) for a single task. The time complexity of this part is also O(1) because the user only considers a small subset of edge servers that fall within the delay threshold when calculating the candidate servers. In conclusion, an individual task's lifetime time complexity is O(m). When the problem size is n, the overall time complexity is O(nm).

9.4 Analysis of the Intelligent Edge Server Selection Method

On a Mac with the MACOS operating system, the SeeMe approach was simulated. The system included a 4-core CPU with a 2.2 GHz primary frequency. Acted out the situation of a wagon carrying goods without a driver. There were several randomly created routes in the scenario. The tasks were generated at random over a certain period of time, and they ended when the unmanned cart that was transporting the supplies came to an end.

Exams conducted in the real world provided the data used in the experiments. For instance, a small edge server typically has between 1 and 3 TB of memor. Abstracted

Table 9.1 Experimental settings

Variable	Value range
The coverage radius of the base station	200 m
Number of edge servers	60
Total system uptime	500 s
The total number of tasks in the mobile track	6000, 8000, 10,000, 12,000
r	[5–10 Mb/s]
Time slot	100 ms

Table 9.2 Resource distribution of edge servers

(a) server index (1–20)		(b) server index (21–40)		(c) server index (41–60)	
Index	Storage(Gb)	Index	Storage(Gb)	Index	Storage(Gb)
1	1618	21	1044	41	1453
2	2092	22	2855	42	1089
3	1907	23	2483	43	1549
4	1298	24	1569	44	1726
5	2317	25	1087	45	1158
6	2318	26	1270	46	1162
7	1392	27	1281	47	1501
8	1382	28	1495	48	2227
9	2067	29	2697	49	2218
10	2162	30	2907	50	2178
11	2764	31	2600	51	1421
12	1120	32	1985	52	1023
13	2532	33	2031	53	1944
14	1695	34	2218	54	2426
15	1737	35	2924	55	2070
16	1211	36	2757	56	1711
17	2004	37	1700	57	1091
18	2653	38	1632	58	2417
19	2489	39	2493	59	2032
20	1368	40	1505	60	2371

the memory size needed for the tasks to run to between 100 MB and 10 GB because a small edge server typically serves 200–1000 tasks. On the basis of an actual survey, additional related data were also generated at random, and the relevant parameters are displayed in Table 9.1. The edge servers in the experiment were all heterogeneous, and each edge server had a different storage configuration to better mimic the real-world scenario, which is shown in Table 9.2.

Four approaches—always-transfer, short-sighted policy, probability-based policy, and MDP-based policy—were chosen for comparison with SeeMe. These methods' secondary optimization goals may overlap, but their main goal is to optimize differently. The always-transfer approach's main optimization goal is to reduce average latency. The short-sighted policy's primary optimization objective is to increase average availability. The probability-based policy's primary optimization objective is to load balance the system, whereas the MDP policy's primary optimization objective is to reduce system overhead. The aforementioned methods are implemented in C++, and the four methods that were compared are briefly described as follows:

Always-transfer (AT): When the base station to which a task is attached changes, the task is immediately moved to the closest edge server. The cloud is used to run jobs that are queued up.

Short-sighted policy (SSP): Active transfer, which moves the task to a close-by edge server, is activated when the task latency hits a threshold. A job is transferred to a nearby edge server that is not overburdened for execution when passive transfer is triggered by the queueing of the task. The cloud is used to run jobs that are queued up.

MDP-based policy (MBP): Tasks are transferred to the first candidate edge server with available resources when transfer is triggered. The MDP model is utilized to determine the best list of candidate edge servers. The cloud is used to run jobs that are queued up.

Transfer is initiated by probability-based policy (PBP): when a task has just been generated or when a delay threshold is achieved. The edge server does a Bernoulli trial to determine the service assignment after the transfer is initiated. In the experiment, the probability distribution-adjustable shape parameter was set to 2. They were moved to run on the cloud for tasks that were queued.

9.4.1 Evaluation of Queued Tasks Number

Consider the 6000, 8000, 10,000, and 12,000 tasks that make up the system's total number of queues. The task queue count describes how many times an edge server overload forces a task to be queued for execution. It is a metric that depicts the system's load status. The algorithm's load-balancing performance improves with the number of task queues. Some user tasks queue up for service on the edge server as a result of resource-blind competition. The total number of queues grows by one when user tasks are queued for this reason. User tasks are moved to the cloud for execution once they reach the queue-waiting state in the experiments instead of actually queuing up on the edge servers. The cumulative queue count gives a good idea of how frequently user tasks compete blindly for resources. It goes without saying that the fewer tasks in the queue, the better.

As the number of user tasks rises, as shown by the experimental results in Fig. 9.3, the cumulative queue counts for all five algorithms increase monotonically. The number of user tasks causing queue-wait conditions rises as user task competition for resources becomes more pronounced. It is clear that for all task densities, the number of SeeMe queues is consistently close to 0. Due to the regulation of the allocation function, which effectively lowers the possibility of offloading tasks to the

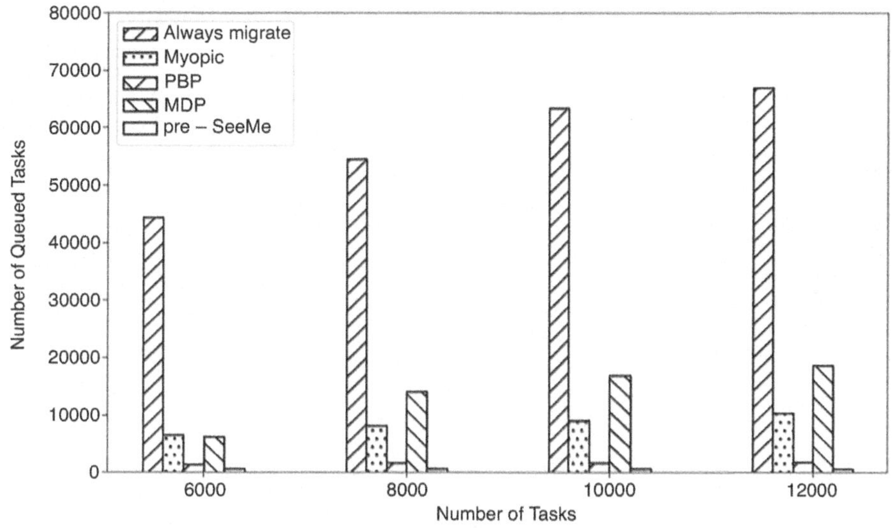

Fig. 9.3 Queued task numbers with different task densities

edge server in high concurrency scenarios, the PBP approach can also effectively reduce the number of queued tasks. Less system overload is indicated by fewer tasks in the queue, which suggests that SeeMe is successfully achieving its goals.

9.4.2 Evaluation of Availability

Availability represents the ratio of the number of tasks served by the edge server to the number of all tasks requesting the edge server's service in each second across the system. This metric gives a general idea of the caliber of the service offered to users. Better algorithms result in lower costs because users intuitively perceive lower latency when their tasks are processed by the edge server, which improves the user experience and increases revenue for the company.

Use average availability to highlight the benefits and drawbacks of these five algorithms because real-time availability within each time slot varies significantly over time. The average availability for each time slot is referred to as average availability. An accurate estimate of the percentage of users the edge server can serve during each time slot can be obtained from average availability. The higher the average availability is, the better the QoS for users.

The average availability of the system within each time slot is illustrated in Fig. 9.4. Overall, the availability of the different methods decreases monotonically as the number of user tasks increases, which is consistent with a realistic pattern. The SSP, MBP, and SeeMe methods all perform better in terms of availability when there are 6000 user tasks, with the SeeMe method essentially achieving the highest

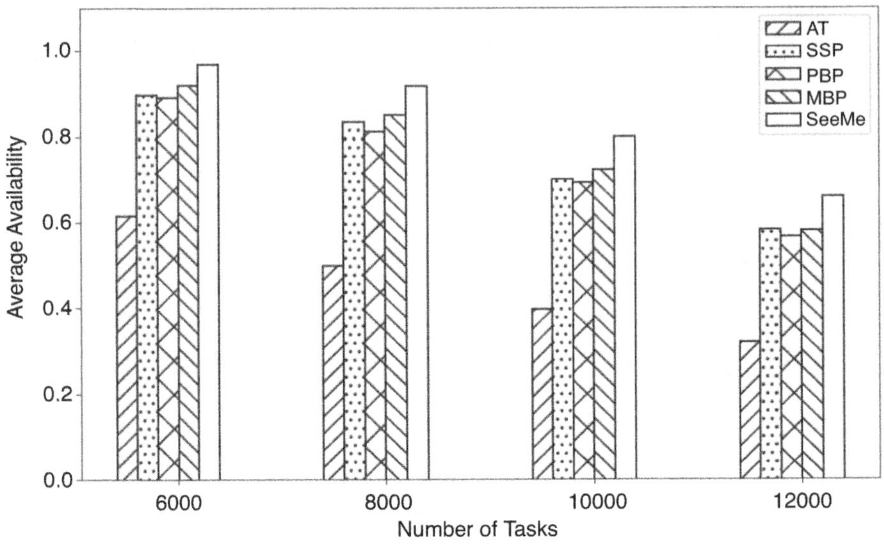

Fig. 9.4 Availability changes with different task densities

average availability value of 1. The average availability of all four methods gradually declines as the number of user tasks rises, but the SeeMe method consistently has the highest average availability and outperforms several other methods by at least a significant margin. The SeeMe method can serve at least 80 more users at the edge server, presuming 10,000 tasks in the system, which is significant in terms of user experience.

9.4.3 Evaluation of System Cost

Considers the difference in system cost between SeeMe and three other methods at medium to high task densities. The loss of service providers is inversely proportional to the system cumulative cost. The 8000, 10,000, and 12,000 tasks' cumulative system costs are depicted in Fig. 9.5's three graphs, which total more than 500 seconds. Because the AT method only takes into account putting tasks on the closest edge server, which means that most tasks cannot be satisfied by the edge and must instead be processed in the cloud, its system cost is always the highest of the four methods. Due to two main factors, the SSP method has the next highest cumulative system overhead. On the one hand, the point at which SSP initiates migration is not the least expensive one, but rather the point at which the delay exceeds a certain amount. However, frequent migrations in scenarios with high concurrency result in excessive migration overheads. Because PBP effectively prevents edge server overload through probabilistic offloading, it is able to solve the issue of frequent passive migrations brought on by SSP. As a result, the overheads of PBP and MBP

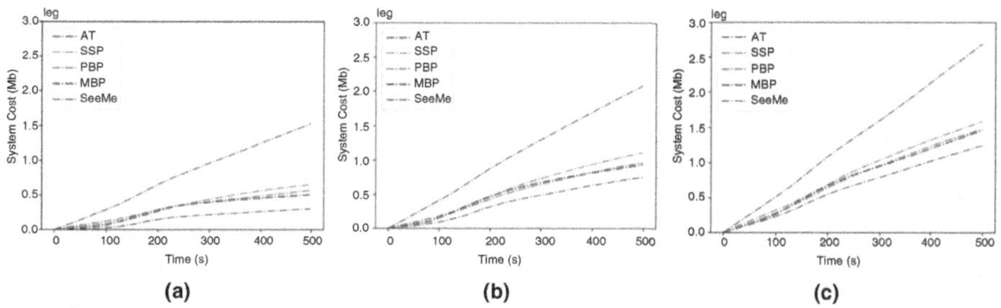

Fig. 9.5 System cost changes with different task densities. (**a**) 8000 tasks. (**b**) 10,000 tasks. (**c**) 12,000 tasks

are comparable. When choosing candidate servers, MBP, in contrast, is a lower-cost priority and initiates migration at a better time. By choosing the least expensive server offload task while avoiding server overload, SeeMe combines the benefits of PBP and MBP to achieve the lowest cumulative system overhead.

Finally, the cumulative system overhead increases along with the total number of users. The percentage reduction for system cost declines, with SeeMe only cutting the cost by compared to MBP for a user count of AT 12,000 users, despite the fact that the SeeMe approach has a bigger absolute difference in system cost compared to the SSP, PBP, and MDP methods. In conclusion, SeeMe reduces system expenses effectively in situations with medium-load to high-load.

9.4.4 Evaluation of Average Communication Hops

Considers the average delay measurement. The number of communication hops between the task and the edge server represents the communication delay, with fewer communication hops indicating a better quality of service (QoS) for the user.

The relationship between the average number of communication hops for a task and the number of tasks is shown in Fig. 9.6. According to the findings, SeeMe has the fewest average communication hops. In particular, the AT method consistently maintains a high average communication hop count due to a significant portion of its tasks being carried out in the cloud. By setting the cloud-to-side communication hop count to 20, this method consistently exhibits the highest average communication hop count. Because all three methods have some drawbacks, the average communication hop counts of SSP, PBP, and MBP are comparable. High concurrency scenarios cause SSP to initiate multiple passive migrations, which restricts tasks' access to services to distant edge servers with free resources. The PBP method effectively avoids the issue of overloading the edge servers because in a high concurrency scenario, the probability of successfully offloading the PBP method tends to zero. In addition, if resource utilization is not close to, more tasks may be carried out in the cloud. The SeeMe method, which combines the benefits of these

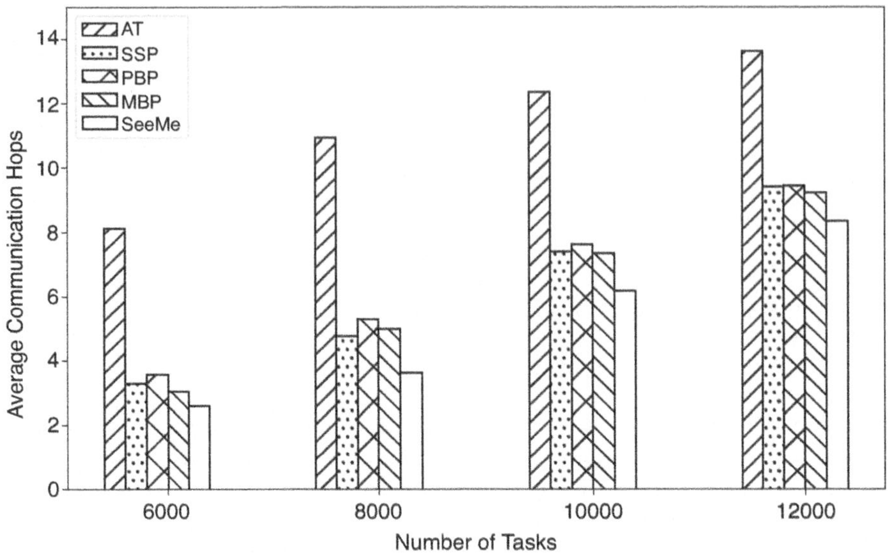

Fig. 9.6 Communication hop changes with different task densities

methods and has the best performance in terms of communication latency, has a lower average hop count than MBP because MBP overloads some of the more heavily loaded edge servers, forcing some of the tasks to be processed in the cloud.

9.5 Summary

This chapter investigates the framework of mobile computing and Industrial Internet applications in managing business tasks. This framework considers the location and mobility of workers in industrial settings, enabling more efficient task management. The SeeMe server selection procedure is then introduced. This approach concentrates on the novel location-aware environment that is presented. The majority of consumers can receive edge service from SeeMe in this setting at the lowest possible system cost. Locking actions during resource distribution can also stop users from receiving restricted resources. Convincing tests were carried out in a multi-mobile working environment, and the outcomes show that SeeMe excels in a number of areas.

References

1. Liu J, Wang X, Shen S, Yue G, Shui Y, Li M (2020) A Bayesian q-learning game for dependable task offloading against DDoS attacks in sensor edge cloud. IEEE Internet Things J 8(9):7546–7561
2. Yan C, Zhang Y, Zhong W, Zhang C, Xin B (2022) A truncated SVD-based ARIMA model for multiple QoS prediction in Mobile edge computing. Tsinghua Sci Technol 27(2):315–324
3. Yuan L, He Q, Chen F, Zhang J, Qi L, Xiaolong X, Xiang Y, Yang Y (2022) CSEdge: enabling collaborative edge storage for multi-access edge computing based on Blockchain. IEEE Trans Parallel Distributed Systms 33(8):1873–1887
4. Xu X, Fang Z, Qi L, Zhang X, He Q, Zhou X (2021) TripRes: traffic flow prediction driven resource reservation for multimedia IoV with edge computing. ACM Trans Multimedia Comput Commun Appl 17(2):1–21
5. Zhao H, Liu Z, Tang J, Gao B, Zhang Y, Qian H, Huaqiang W (2022) Memristor-based signal processing for edge computing. Tsinghua Sci Technol 27(3):455–471
6. Wang F, Wang L, Li G, Wang Y, Lv C, Qi L (2021) Edge-cloud-enabled matrix factorization for diversified APIs recommendation in mashup creation. World Wide Web 24:1–21
7. Shen S, Huang L, Zhou H, Shui Y, Fan E, Cao Q (2018) Multistage signaling game-based optimal detection strategies for suppressing malware diffusion in fog-cloud-based IoT networks. IEEE Internet Things J 5(2):1043–1054
8. Xu X, Tian H, Zhang X, Qi L, He Q, Dou W (2022) DisCOV: distributed COVID-19 detection on x-ray images with edge-cloud collaboration. IEEE Trans Serv Comput 15(3):1206–1219
9. Shen Y, Shen S, Zongda W, Zhou H, Shui Y (2022) Signaling game-based availability assessment for edge computing-assisted IoT systems with malware dissemination. J Inf Secur Appl 66:103140
10. Shi T, Cai Z, Li J, Gao H (2020) CROSS: a crowdsourcing based sub-servers selection framework in D2D enhanced MEC architecture. In: 2020 IEEE 40th International Conference on Distributed Computing Systems (ICDCS), pp 1134–1144
11. Tu H, Xu H, Huang L, Yang X, Yao D (2020) Joint server selection and SFC routing for anycast in NFV-enabled SDNs. In: International Conference on Wireless Algorithms, Systems, Applications, pp 432–443
12. Wu Z, Li G, Shen S, Lian X, Chen E, Guandong X (2021) Constructing dummy query sequences to protect location privacy and query privacy in location-based services. World Wide Web 24(1):25–49
13. Bhooanusas N, Sok-Ian Sou (2021) Measurement-basediot server selection for mobile edge computing. In: 22nd Asia-Pacific Network Operations; Management Symposium (APNOMS), pp 17–20
14. Ho TM, Nguyen K-K (2021) Deep q-learning for joint server selection, offloading, and handover in multi-access edge computing. In: ICC 2021-IEEE International Conference on Communications, pp 1–6
15. Ding Y, Liu C, Li K, Tang Z, Li K (2019) Task offloading and service migration strategies for user equipments with mobility consideration in mobile edge computing. In: 2019 IEEE Intl Conf on Parallel and Distributed Processing with Applications, Big Data and Cloud Computing, Sustainable Computing and Communications, Social Computing and Networking (ISPA/BDCloud/SocialCom/SustainCom), pp 176–183
16. Yang J, Han J, Xing Y, Zhang Y, Wei W, Xue K (2020) SSMP: server selection for multipath TCP in CDN environments. In: IEEE Global Communications Conference, pp 1–6

Chapter 10
Intelligent Business Resources Deployment Over Industrial Internet

Abstract Business process management (BPM) is a crucial method for standardized and systematic management in the context of Industrial Internet, especially Industrial Internet of Things (IIoT). As the mobile internet develops rapidly and the volume of data generated by IoT devices increases, traditional storage schemes are no longer sufficient for managing mobile IIoT business data. To address this challenge, the cloud-edge collaborative data storage scheme has emerged as an effective solution. This approach leverages cloud data centers to provide mass storage capability, while also considering access delays by storing data on edge servers located near users. Given the constrained storage capacity of edge servers, the development of an efficient data placement strategy has emerged as a critical issue in enhancing the speed of data retrieval and task execution within IIoT business processes.

In this chapter, we will explore the use of Bayesian optimization algorithm (BOA) in the context of IIoT resource placement strategy. Resource deployment plays a crucial role in optimizing business processes and achieving efficient utilization of resources. To enhance user experience, shorten overall running time, and reduce data transmission costs, a data placement strategy based on Bayesian optimization algorithm is discussed in this chapter. This strategy can allocate storage capacity on the edge side in a reasonable manner and determine the most appropriate edge server on which to place the data. The simulation results show that the BOA-based strategy outperforms those based on other optimization algorithms used in similar studies across varied scenarios.

Business process management (BPM) is a crucial method for standardized and systematic management in the context of Industrial Internet, especially Industrial Internet of Things (IIoT). As the mobile internet develops rapidly and the volume of data generated by IoT devices increases, traditional storage schemes are no longer sufficient for managing mobile IIoT business data. To address this challenge, the cloud-edge collaborative data storage scheme has emerged as an effective solution. This approach leverages cloud data centers to provide mass storage capability, while also considering access delays by storing data on edge servers located near users. Given the constrained storage capacity of edge servers, the development of an

W. Dou et al., *Intelligent Industrial Internet Systems*,
https://doi.org/10.1007/978-981-99-5732-3_10

efficient data placement strategy has emerged as a critical issue in enhancing the speed of data retrieval and task execution within IIoT business processes.

In this chapter, we will explore the use of Bayesian optimization algorithm (BOA) in the context of IIoT resource placement strategy. Resource deployment plays a crucial role in optimizing business processes and achieving efficient utilization of resources. To enhance user experience, shorten overall running time, and reduce data transmission costs, a data placement strategy based on Bayesian optimization algorithm is discussed in this chapter. This strategy can allocate storage capacity on the edge side in a reasonable manner and determine the most appropriate edge server on which to place the data. The simulation results show that the BOA-based strategy outperforms those based on other optimization algorithms used in similar studies across varied scenarios.

10.1 Resource Deployment for Business Process Management

The integration of intelligent Internet of Things (IoT) devices, including temperature sensors, electronic monitors, and radio frequency identification (RFID) tags, within the life cycle of industrial products has given rise to the paradigm of the Industrial Internet of Things (IIoT). Consequently, dynamic information networks have been established at both local and global levels, offering distinctive advantages in terms of enhancing production efficiency, minimizing errors, and bolstering security [1, 2]. Complex IIoT applications often comprise multiple sub-activities that are interconnected through various relationships such as dependencies, parallelism, and mutual exclusion. These relationships are typically described using workflow models or higher-level process models, such as the business process modeling notation (BPMN) [3]. In this chapter, we refer to these applications as Industrial Internet of Things business processes, which are extensively utilized in various domains, including optimization of manufacturing processes and other areas [4].

Resource deployment is a critical technology for business process management (BPM) in IIoT, which involves the allocation of resources, such as computing resources, communication resources, and mobile business data resources, etc., to different tasks or activities in a business process. With the advancement of IoT technology, the volume of data generated by IIoT devices has witnessed explosive growth. This has led to the emergence of data-intensive IIoT business process, which heavily relies on data from widely deployed IoT sensors [5]. However, the devices that request and use these data are often not fixed, and the network topology may change frequently, leading to what we term as IIoT mobile business data. The ongoing expansion of data volume presents new challenges in the management of IIoT mobile business data.

To leverage the benefits of both cloud computing and edge computing, the paradigm of cloud-edge collaboration has been proposed as a means to enhance

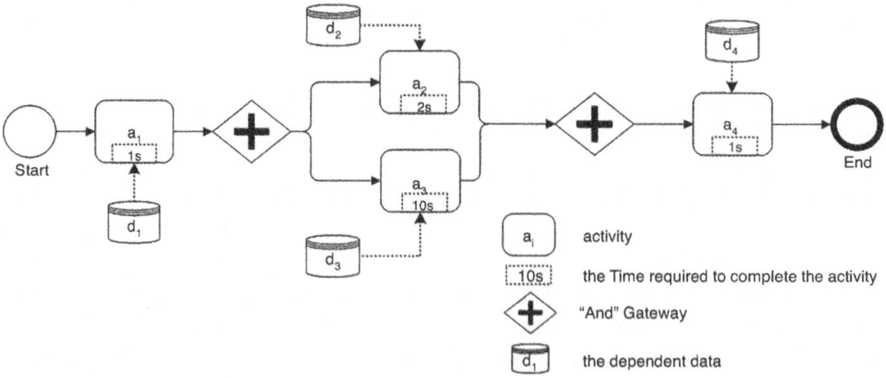

Fig. 10.1 An example of IIoT business process including concurrent sub-activities

the execution efficiency of IIoT business processes and ensure efficient data access [6]. It provides a framework to effectively address the management of IIoT mobile business data. Analyzing the architecture of the cloud-edge collaboration environment from the perspective of data storage, the cloud data center can offer massive and scalable storage capabilities, while the edge server (ES) provides limited capacity but low access latency. Hence, determining what type of data should be placed on the edge side and which ES these data should be placed on plays a critical role in improving access efficiency. This challenge is referred to as the IIoT mobile business data placement problem. An optimal data placement strategy can accelerate data transmission, reduce the cost of data transmission, and consequently impact the overall running time and cost of the entire IIoT business process [7].

In the subsequent part of this section, we will provide formal definitions of IIoT business process, IIoT mobile business data, and edge-cloud collaborative environment. Furthermore, we will delve into the IIoT mobile business data placement problem to provide a comprehensive understanding of the topic.

IIoT Business Process. An IIoT business process is composed of various sub-activities. Each execution of the process can be considered as a trace of these sub-activities, and a collection of such trace forms an events log. We define an event log as T activity traces, where each trace comprises T_i $(i = 1, 2, \ldots, T)$ sub-activities, represented as $\text{Trace}_i = < a'_1, a'_2, \ldots, a'_{T_i} >$. Here, a'_i is a sub-activity that belongs to the activity set $A = \{a_1, a_2, \ldots, a_N\}$, where N denotes the number of activity types. In contrast to the conventional scientific workflow calculation, the order and frequency of activities in each activity trace may differ. For instance, in the BPMN process model depicted in Fig.10.1, both $<a_1, a_2, a_3, a_4>$ and $<a_1, a_3, a_2, a_4>$ are valid activity traces. Thus, our focus should be on analyzing the events log that contains numerous activity trace details, instead of a static process model.

IIoT Mobile Business Data. Given that the target business process involves K types of data, each type of business data can be defined as a set $\text{Data} = < d_1, d_2, \ldots, d_K >$. The data dependencies between sub-activities can be represented using a matrix W_{N*K}. Each value of the matrix is defined as:

$$W_{ij} = \begin{cases} x = \text{if } a_i \text{ depends on } d_j, \text{and } x > 0 \\ 0 = \text{if } a_i \text{ does not depend on } d_j \end{cases} \tag{10.1}$$

Here, x is a constant representing the average size of data d_j that is requested by a_i each time. Based on x, we can construct a random variable to estimate the transmission time and cost required to read the corresponding data. Figure 10.1 illustrates the data dependencies between activities and the four different types of data, where d_1 is read when a_1 is executed, and d_2 is read when a_2 is executed. Similarly, the dependencies between other types of data and activities can be inferred.

Edge-Cloud Collaborative Environment. This chapter primarily focuses on the storage capabilities within an edge-cloud collaborative environment. The cloud data center is denoted as D_c and is assumed to have unlimited data capacity. The edge side storage cluster is represented as $E = \ < E_1, E_2, \ldots, E_M>$, where M denotes the number of edge side servers with varying storage capacities and transmission abilities such as bandwidth and delay. Each E_i is defined as a triplet comprising the maximum data capacity, communication bandwidth with end users and transmission delay respectively, as illustrated as follows,

$$E_i = \ < \text{Cap}_i, \text{Band}_i, \text{Delay}_i > \tag{10.2}$$

IIoT Mobile Business Data Placement Problem. Data placement has long been a focal point of research in computer science. Initially, research efforts were largely centered on optimizing data placement strategies for storage components such as disks and memory within a single computer system. However, with the advent of big data and the emergence of various architectures such as cloud computing and edge computing, optimizing data placement strategies in distributed systems has become a critical factor in reducing data transmission costs and enhancing operational efficiency.

Conventional data placement strategies have traditionally relied on clustering algorithms [8] or evolutionary algorithms [9, 10] to address the needs of scientific computing scenarios [8, 11, 12], where the size of the dataset is generally static and does not account for the ongoing generation of data by widely deployed devices like IoT sensors. In addition, previous research has tended to prioritize the minimization of data transmission time. However, in IIoT business processes that encompass parallel sub-activities, a sole focus on data transmission time may not provide a comprehensive solution, as it fails to account for waiting events. Figure 10.1 illustrates an IIoT business process using a simple BPMN model, where the dotted boxes indicate the computation time required for each sub-activity (a_1 to a_4). In the simplest case, the computation time of each sub-activity is significantly larger than the data transmission time, and each sub-activity necessitates the same amount of data to be read during its execution. It should be noted that the data type required for each sub-activity is distinct (d_1 to d_4, respectively), and considering the limited storage capacity of the edge side, only two data types can be accommodated on the edge side. Therefore, in terms of data transmission time, it is inconsequential that two data types are selected for placement on the edge side, as it does not impact the overall data transmission time. Nonetheless, in practice, the placement of data type d_2, as required by a_2, on the edge side does not significantly reduce the overall running time of the business process, given that its parallel task a_3 has a considerably longer task execution time. Thus, the acceleration of data loading achieved by a_2 on the edge side does not effectively contribute to the running time.

The aforementioned example demonstrates the impact of parallel sub-activities. Given the intricacy of IIoT business processes and the influence of other factors such as computational offloading [13], it is sensible to infer that optimizing solely for data transmission time may not be all-encompassing. Therefore, the subsequent section shall introduce a novel approach to include the overall execution time of the IIoT business process as an additional optimization objective.

10.2 Intelligent Business Resource Deployment Based on Bayesian Optimization

The Bayesian optimization algorithm offers distinct advantages in addressing black-box optimization problems such as hyperparameter tuning in deep learning tasks. In addition, given the mobile nature of IIoT network topology, the strategy must be updated regularly. Traditional evolutionary algorithms such as genetic algorithms (GAs) [14] and particle swarm optimization (PSO) [15] algorithms need numerous evaluations of the objective function to attain the desired outcome and may not be suitable for the frequent update scenario. Conversely, BOA requires fewer evaluations when solving, making it more suited to this scenario [16]. Consequently, this section introduces a novel data placement strategy for IIoT mobile business based on BOA that considers the overall IIoT business process's runtime and effectively enhances the computational efficiency and actual data transmission performance [17].

We will present formal definitions of both the data placement strategy and optimization objective, followed by a detailed exploration of the Bayesian optimization algorithm.

Data Placement Strategy. The concept of a data placement strategy can be formulated as follows:

$$\text{Strategy} = <(\text{size}_1, \ldots, \text{size}_K), (\text{loc}_1, \ldots, \text{loc}_K)> \tag{10.3}$$

It encompasses two essential components that govern the allocation of the ith category of data in the corresponding edge location loc_i and restrict it to a maximum size of size_i. To provide a clearer explanation, we can delineate the two components of the strategy as $V_s = \ < s_1, s_2, \ldots, s_k>$, which denotes the maximum data size permitted in the edge, and $V_p = \ < p_1, p_2, \ldots, p_k>$, which represents the placement location for the data. Therefore, we can represent the strategy as Strategy $= \ < V_s, V_p>$.

Optimization Objective. The objective of this strategy is to reduce the overall running time of the entire IIoT business process by proposing a suitable data placement strategy that reduces the cost of data transmission. The running time of an IIoT business process is affected by various factors, including device computing capability, data transmission time, and I/O waiting time. This strategy specifically focuses on the impact of data placement on the overall running time, as the data placement strategy directly affects data transmission time. To achieve the optimization objective, this strategy considers the overall running time, as optimizing only for data transmission time may not lead to satisfactory results due to sub-activity constraints and computational offloading [18].

If prior knowledge on the average computation time (represented by t_i, excluding data transmission time) and running location (represented by w_i, with values of local, edge, and cloud representing computing tasks offloaded to the corresponding area) of each activity is available from historical data, the running time rt_i to complete an activity can be calculated as follows:

$$rt_i = \begin{cases} t_i + p * \left(c_1 * \dfrac{\mathrm{Req}_i}{\mathrm{Band}_{\mathrm{edge}}} + \tau_e\right) + (1-p) * \left(c_2 * \dfrac{\mathrm{Req}_i}{\mathrm{Band}_{\mathrm{cloud}}} + \tau_c\right) & \text{if } w_i = \text{Local} \\ t_i + (1-p) * \left(c_1 * \dfrac{\mathrm{Req}_i}{\mathrm{Band}_{\mathrm{cloud}}} + \tau_c\right) & \text{if } w_i = \text{Edge} \\ t_i & \text{if } w_i = \text{Cloud} \end{cases} \tag{10.4}$$

Here, p is a variable with a value range of 0 or 1, depending on whether the data is located on the edge side or not [19]. $\mathrm{Band}_{\mathrm{edge}}$ and $\mathrm{Band}_{\mathrm{cloud}}$ are the corresponding transmission bandwidths of the server where the data is located, whereas τ_e and τ_c are the transmission delays of the edge servers and cloud servers, respectively. We assume that the cloud data center contains all backup data and ignore the data transfer time between the same data center. In addition, to simulate the actual operation of the IIoT business process and train the data placement strategy, we refer to this equation to build an environment.

The overall running time $G(\mathrm{Tr})$ required by an active trace Tr can be obtained with Algorithm 10.1.

Algorithm 10.1: Calculate the Running Time of an Activity Trace

1: **Input:** an activity trace *Tr*
2: **Output:** $G(Tr)$, which is the run time of *Tr*
3: Initialize the start time $t_{start} = 0$, a set of completed activity $D = \{\}$, an empty dict of each activity completed time *Dt*
4: **for** each activity *a* in *T* **do**
5: **if** *a* does not depend on other activities **then**

(continued)

```
        t_a ← rt_i        // calculate by (8)
    else
        D'← the dependent activities
        if D'∈ D then
            t_e ← the max completed time in D'
            x ← rt_i // the execute time
            t_a ← t_e + x // the completed time
        else
            skip to next iteration
        end if
    end if
    D ← D∪{a}
    Dt[a] ← t_a
    T ← T / {a}
    if T is empty then
        stop the loop
    end if
end for
return max(Dt.values())
```

During strategy training, event logs consisting of many traces (denoted as DS) are typically used as inputs, and the running time of the entire event logs F(DS) can be expressed as follows:

$$F(\mathrm{DS}) = \sum_{i=1}^{T} G(\mathrm{Trace}_i) \tag{10.5}$$

The optimization objective can be defined as follows:

$$\begin{gathered} \min F(\mathrm{DS}) \\ s.t. \sum_{i=1,\, \mathrm{loc}_i = j}^{K} \mathrm{size}_i \le \mathrm{Cap}_j, \mathrm{j} = 1, 2, \ldots, M \end{gathered} \tag{10.6}$$

Here, the optimization goal is the overall running time of the entire IIoT business process, and the constraints indicate that the amount of data placed on each edge server cannot exceed the server's capacity limit. F(DS) depends on certain key prior knowledge, including running time and location, which may be missing or incomplete. In such cases, we can simulate the real scenario multiple times, which can be time-consuming and resource-intensive. This is one of the main reasons why we choose the BOA as the training method.

Bayesian optimization algorithm (BOA). BOA is a probabilistic surrogate model-based optimization algorithm that has several advantages over evolutionary algorithms, including its enhanced interpretability [20]. BOA is typically used to solve combinatorial optimization problems with a high computational cost for the objective function. Additional benefits of utilizing BOA include its ability to find the global optimal solution, require fewer iterations to obtain an acceptable solution, and improve interpretability [21]. The primary objective of BOA is to solve problems represented as follows:

$$\min f(x), x \in R^d \tag{10.7}$$

Here, $f(x)$ is the objective function. When applying BOA, knowledge of the algebraic expression of $f(x)$ or its derivatives is unnecessary. However, it is essential to ensure that $f(x)$ is continuous or nearly continuous to use Gaussian process modeling. The input vector is represented by x, while d represents the dimension of the input vector. It is generally better to have a lower value of d as BOA performs better in search spaces of lower dimensions [22]. The primary approach of BOA is to identify the most valuable point to evaluate and update the posterior information according to the previous knowledge of the function f to maximize the model's accuracy [23].

The design of the surrogate model and acquisition function is considered the most critical aspect of BOA. The surrogate model is a Bayesian statistical model used to model the objective function, while the acquisition function determines the next group of sampling positions. Gaussian process is commonly used as a surrogate model, where any finite random variable satisfies a joint Gaussian distribution:

$$f(x) \sim \mathcal{GP}(m(x), k(x, x')) \tag{10.8}$$

Here, the mean value function, $m(x) = E[f(x)]$, and the covariance function, $k(x, x') = E[(f(x) - m(x))(f(x') - m(x'))]$, are critical components of the surrogate model. The role of the surrogate model is to restore the information of the objective function as much as possible based on the limited sampling points. The Gaussian process provides mean and variance estimations for each unknown point.

The acquisition function determines the optimal sampling point and significantly influences whether BOA can converge correctly to the optimal point. A good acquisition function should consider both the exploitation of known points and the exploration of unknown points. Commonly used acquisition functions include the probability of improvement (PI) function and the expected improvement (EI) function. PI and EI can be calculated as follows:

$$\mathrm{PI}(x) = p(f(x) \le f(x_{\text{best}}) - \xi) = \phi\left(\frac{f(x_{\text{best}}) - \xi - \mu_t(x)}{\sigma_t(x)}\right) \tag{10.9}$$

$$\mathrm{EI}(x) = (f(x_{\text{best}}) - \mu_t(x))\phi\left(\frac{f(x_{\text{best}}) - \mu_t(x)}{\sigma_t(x)}\right) + \sigma_t(x)\phi\left(\frac{f(x_{\text{best}}) - \mu_t(x)}{\sigma_t(x)}\right) \tag{10.10}$$

where x_{best} represents the current optimal point, while ξ controls the threshold value of promotion. PI only focuses on the probability of promotion, which is considered insufficient for the unknown point's promotion. To address this issue, the EI function focuses on the expectation of the unknown point's promotion.

10.3 BOA-Based Business Resource Deployment over Industrial Internet

In this section, we will discuss how BOA can be used in the business resource deployment for the Industrial Internet to reduce overall running time [24]. In addition, we simulate the data placement problem for data-intensive IIoT business processes in a cloud-edge collaboration environment and demonstrate the effectiveness of BOA-based business resource deployment.

IIoT mobile business data placement strategy based on BOA. As mentioned above, a data placement strategy can be expressed as V_s and V_p. We can divide this problem into two smaller sub-problems for more effective resolution. Given that various edge servers (ESs) have different storage capacities, their end-to-end data transmission capabilities, including bandwidth and delay, are quite similar and generally better than cloud-to-end data transmission capabilities. Consequently, it is entirely feasible to break down this data storage problem into two stages to solve it. As the probability of locating requested data at the edge is significantly influenced by the size of the corresponding data type, solving V_s takes priority over V_p. This means that we first allocate data size, then determine data location, and finally combine them to create a complete data placement strategy. When addressing V_s, we use the average edge-to-end delay and bandwidth of each ES as inputs to the objective function for computation and obtain an estimate. To solve V_s, we introduce the actual delay and bandwidth of each ES to calculate the final real overall runtime.

To enhance the accuracy of results, this strategy has made improvements to the BOA, with a focus on redesigning the acquisition algorithm. Generally, the acquisition function must consider both exploration and exploitation. The commonly used PI and EI acquisition functions can adjust their corresponding parameters to emphasize making the most of the current optimal point or exploring other unknown regions. In this section, we adopt the EI function as the basic acquisition function, which considers the degree of improvement and has proven to be efficient in many situations. To address the issue of computational efficiency, we choose to search within the rectangular domain of the optimization process, which may lead to a point that violates the capacity limit and renders the EI function invalid.

To solve this problem and enhance the search efficiency of BOA, we have designed a new acquisition algorithm, as shown in Algorithm 10.2.

Algorithm 10.2: The Improved Acquisition Method

1: **Input:** The set of current sample points set X, the size of optimization set u, the adjust function ϕ
2: **Output:** the next selected point x_{next}
3: Initialize a vector V of size u if not exists// velocity vector
4: Sample u points from X to initialize *Pop*
    if not exists// population vector
5: Initialize a vector H as the history optimal
    point of each element in *Pop*
6: *Pop'*← the points have not been evaluated

(continued)

```
if Pop' is not empty then
      x_l ← sample a point from Pop'
      return x_l
end if
x_2 ← EI ()
if x_2 satisfies the capacity constraint then
      return x_2 // as the next point to be evaluated
else
      for i in range(0, u) do
            V_i ← V_i + r_1(x_i − x) + r_2(x_i − H_i)
            Pop_i ← φ(Pop_i + V_i * r_3)
            if Pop_i's performance proved to be better then
                  H_i ← Pop_i
      end if
   end for
end if
```

By introducing a velocity vector and searching for unknown points based on the global optimal point and individual optimal point, inspired by the PSO algorithm, we can search the unknown region more efficiently. We also introduce an adjustment function to handle points that violate the capacity constraint after updating. The adjustment function is not the same for solving V_s and V_p, and we will explain this in detail below.

When solving V_s, the constraints are relatively simple, as shown in the following:

$$\sum_{i=1}^{K} Vs_i \leq \sum_{i=1}^{M} \mathrm{Cap_i} \tag{10.11}$$

Note that the constraint is an inequality constraint. Consider a solution vector $V_s = <s_1, s_2, \ldots, s_k>$ that violates the capacity constraint stipulated in this equation. We define following equation with C being the total storage capacity of the edge side as the adjustment function passed to Algorithm 10.2 to adjust the solution,

$$s_i'' = C * \frac{s_i'}{\sum_{i=1}^{K} s_i'} \tag{10.12}$$

When solving V_p, the problem is that there are K types of constraints, and some of the variables are discrete instead of continuous. This means that we need to handle the relevant hyperparameters carefully to avoid falling into local optima and making invalid explorations in regions where constraints are violated. For the velocity vector V, we stipulate that its value can only be an integer and the range is $[1, M]$ in Algorithm 10.2 when solving V_p. Regarding the adjustment function, we adopt the following strategies. For each server E that exceeds capacity, we sort out the data type in ES by the corresponding data size from small to large and place them in other servers with sufficient capacity using the minimum adaptation policy until the capacity constraint is met in ES again. After this process, there may be some types of data left that have no ES to place, and for each of them, we always choose the server that has the maximum capacity to place the data even if it violates the capacity constraint. Finally, we call the above equation to ensure the capacity constraint is met again, which will change the content of V_s.

After solving V_s and V_p using BOA with our improved acquisition algorithm, the data placement strategy can be expressed as $<V_s, V_p>$.

Fig. 10.2 depicts the management of data derived from Industrial Internet of Things devices by V_s and V_p, which is commonly referred to as the data placement strategy. The strategy can be described as follows:

1. Data of this kind is stored in the ES specified by V_p if it does not exceed the edge side capacity limit defined by V_s.
2. If the capacity limit specified by V_s is surpassed by placing data on the edge side, a portion of old data is removed based on a pre-defined cache management strategy, after which the data is placed in the designated ES as per V_p.
3. When the update interval is reached, the training process is triggered to update the data placement strategy.

Simulation results in IIoT business process. The present study utilizes Python and Linux to simulate the data placement problem that arises in data-intensive IIoT business processes within a cloud-edge collaborative environment. The evaluation criterion used in this study is the average execution time of a single trace consisting of multiple sub-activities. Subsequently, simulation tests are conducted, and the results accurately demonstrate the effectiveness of the proposed data placement strategy. The dataset utilized in the study is a composite of three real workflows, namely Montage, CyberShake, and LIGO [25]. From this composite dataset, a model is extracted and a dataset is constructed, including these features, to simulate an IIoT business process. Table 10.1 presents some crucial parameters and their values that are pertinent to our experiment.

To compare our strategy based on BOA, we selected three heuristic algorithms: DE-DPSO, GA-DPSO, and a stochastic algorithm. These algorithms, namely differential evolution (DE), genetic algorithm (GA), and discrete particle swarm optimization (DPSO), are typical representatives of evolutionary and swarm intelligence algorithms used to solve optimization problems. Combining different heuristic algorithms is a common approach to further enhance their performance. DE-DPSO and GA-DPSO apply the concepts of DE and GA, respectively, to the DPSO algorithm, resulting in good results as reported in the literature. We implemented these algorithms and used them as a comparison with our proposed strategy in the scenario designed in this chapter. We also included a stochastic algorithm as a baseline comparison, ensuring that the capacity constraints were satisfied and the size and location of the data stored on the edge side were completely random. We conducted four different experiments to evaluate the performance of these algorithms under different conditions, while keeping the environment parameters constant except for the independent variables in each experiment.

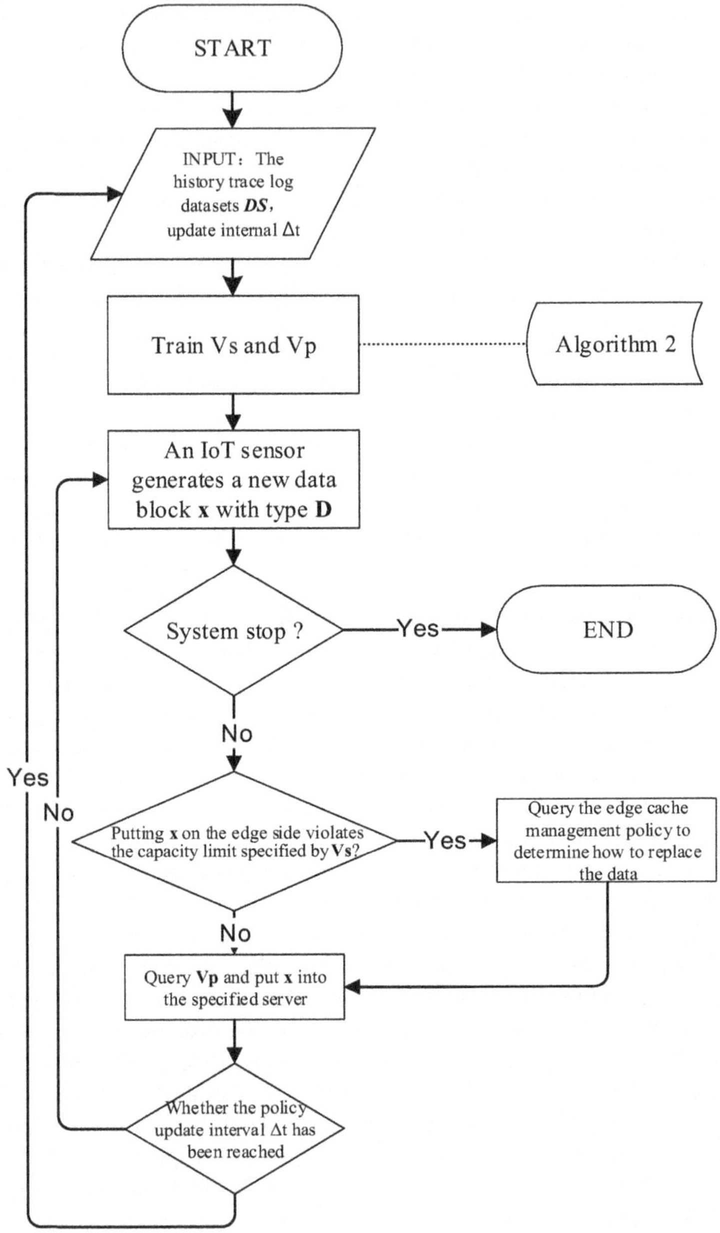

Fig. 10.2 Workflow of IIoT mobile business data placement strategy

Table 10.1 Parameter settings

Parameter	Value
The number of ESs	20
The bandwidth of ESs	50–200 (Mbps)
The bandwidth of cloud servers	10 (Mbps)
The delay of ESs	100–300 (ms)
The delay of cloud servers	1000 (ms)
The number of data types	16
The number of activity types	10
The size of events log	100

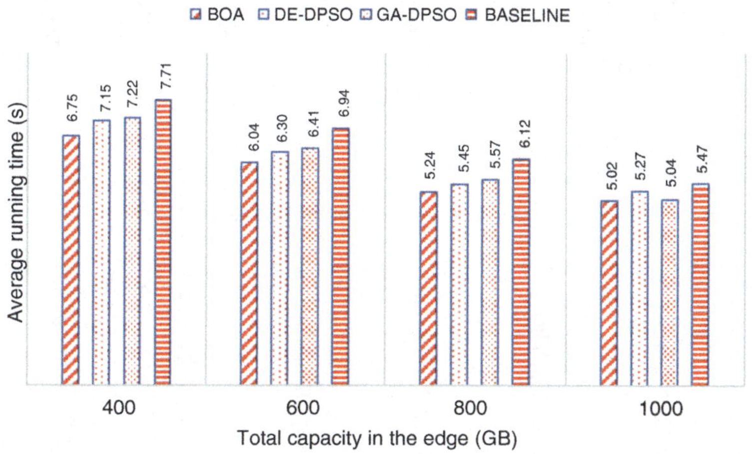

Fig. 10.3 The effect of capacity

Figure 10.3 depicts the impact of the total storage capacity of all energy storages (ESs) on the average running time. The trend on the horizontal axis indicates that for the same algorithm, increasing the capacity of the edge side leads to greater performance improvement. This is because a larger capacity increases the likelihood of data hits on the edge side, thus reducing the data request time. Although the BOA consistently outperforms other algorithms, the performance gap between them decreases as the edge side capacity increases, which is reasonable. As the edge side capacity approaches infinity, the execution time will converge to the same value regardless of the algorithm.

Figures 10.4 and 10.5 illustrate the effect of average edge bandwidth and the number of traces, respectively. Both factors have a significant impact on the algorithms, and the BOA consistently outperforms the other two optimization algorithms and particularly the stochastic algorithm.

Figure 10.6 displays the performance of different algorithms for a fixed number of evaluations. When the number of evaluations is small, the BOA-based algorithm still exhibits superior performance. As the number of evaluations increases, the

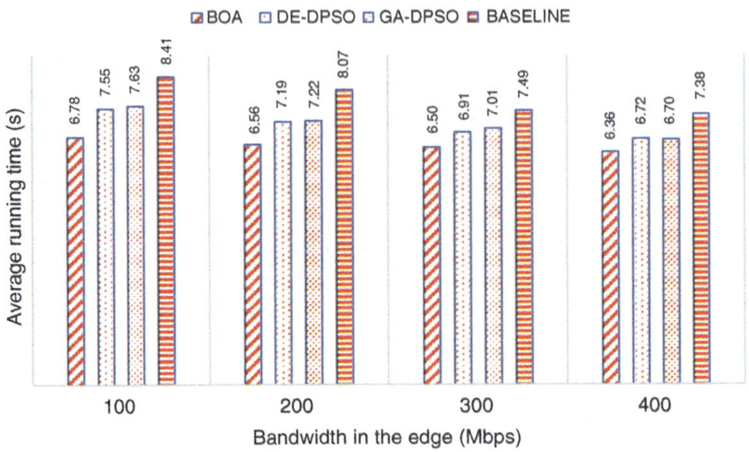

Fig. 10.4 The effect of bandwidth

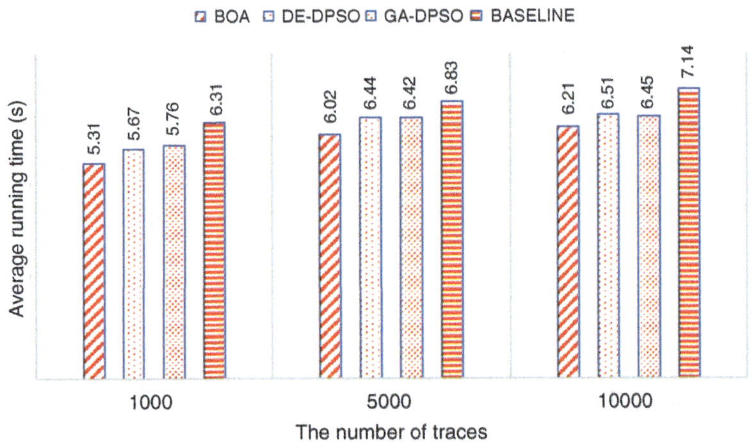

Fig. 10.5 The effect of traces number

performance of all three algorithms improves except the stochastic algorithm, but the BOA-based algorithm remains the best among them.

10.4 Summary

With the exponential growth of data in the realm of IIoT, an increasing number of IIoT business processes are being managed in a collaborative cloud-edge environment. In this chapter, we have discussed the use of Bayesian optimization algorithm for mobile business data deployment in the context of business process management

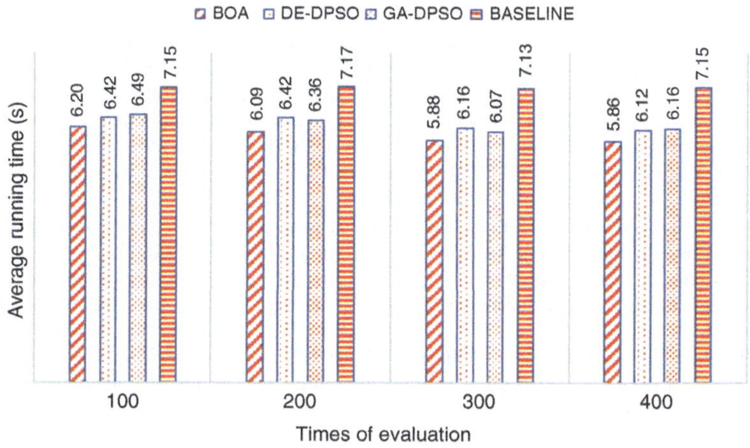

Fig. 10.6 The effect of evaluation times

and IIoT. Mobile business data related to these processes are also being stored on the edge side to expedite access. The BOA-based data placement strategy aims to reduce the overall running time of IIoT business processes and improve their execution efficiency. Experimental results indicate the effectiveness of the BOA-based strategy. In addition, the use of BOA for resource deployment in the context of BPM and IIoT holds great potential in optimizing business processes and achieving efficient utilization of resources. Further research and development in this area can contribute to advancements in the field of IIoT and business process management, leading to improved operational efficiency and enhanced decision-making capabilities for businesses.

References

1. Peng K, Huang H, Zhao B, Jolfaei A, Xu X, Bilal M (2022) Intelligent computation offloading and resource allocation in IIoT with end-edge-cloud computing using NSGA-III. IEEE Transactions on Network Science and Engineering
2. Hazra A, Adhikari M, Amgoth T, Srirama SN (2021) A comprehensive survey on interoperability for iiot: taxonomy, standards, and future directions. ACM Comput Surv (CSUR) 55(1): 1–35
3. Mennuto P, Belahonia JCM, Bazán P (2021) Bpm and socialization tools integrated to improve acquisition and management of information during design and execution of business processes: Bpm-social tool: a proposal. J Comput Sci Technol 21(1):e7
4. Vukšić VB, Stjepić AM, Vugec DS (2021) The intersection between ´ IoT and bpm: Systematic literature review. Proc FEB Zagreb Int Odyssey Conf Econ Business 3(1):824–836
5. Ren L, Liu Y, Wang X, Lü J, Deen MJ (2020) Cloud-edge-based lightweight temporal convolutional networks for remaining useful life prediction in IIoT. IEEE Internet Things J 8(16):12578–12587

6. Wang X, Ren L, Yuan R, Yang LT, Deen MJ (2022) Qtt-dlstm: a cloud-edge-aided distributed LSTM for cyber-physical-social big data. IEEE Transactions on Neural Networks and Learning Systems, (Early Access). https://doi.org/10.1109/TNNLS.2022.3140238
7. Du X, Tang S, Lu Z, Wet J, Gai K, Hung PC (2020) A novel data placement strategy for data-sharing scientific workflows in heterogeneous edge-cloud computing environments. In: 2020 IEEE International Conference on Web Services (ICWS). IEEE, pp 498–507
8. Yuan D, Yang Y, Liu X, Chen J (2010) A data placement strategy in scientific cloud workflows. Futur Gener Comput Syst 26(8):1200–1214
9. Zheng P, Cui L-Z, Wang H-Y, Xu M (2010) A data placement strategy for data-intensive applications in cloud. Jisuanji Xuebao Chin J Comput 33(8):1472–1480
10. Cui L, Zhang J, Yue L, Shi Y, Li H, Yuan D (2015) A genetic algorithm based data replica placement strategy for scientific applications in clouds. IEEE Trans Serv Comput 11(4): 727–739
11. Lin B, Zhu F, Zhang J, Chen J, Chen X, Xiong NN, Mauri JL (2019) A time-driven data placement strategy for a scientific workflow combining edge computing and cloud computing. IEEE Trans Industr Inform 15(7):4254–4265
12. Chen Z, Hu J, Min G, Chen X (2021) Effective data placement for scientific workflows in mobile edge computing using genetic particle swarm optimization. Concurr Computat Pract Exp 33(8):e5413
13. Kang Y, Yang X, Pu B, Wang X, Wang H, Xu Y, Wang P (2022) Hwoa: an intelligent hybrid whale optimization algorithm for multi-objective task selection strategy in edge cloud computing system. World Wide Web 25:1–31
14. Katoch S, Chauhan SS, Kumar V (2021) A review on genetic algorithm: past, present, and future. Multimed Tools Appl 80(5):8091–8126
15. Pace F, Santilano A, Godio A (2021) A review of geophysical modeling based on particle swarm optimization. Surv Geophys 42(3):505–549
16. Frazier PI (2018) A tutorial on bayesian optimization. arXiv preprint arXiv:1807.02811
17. Victoria AH, Maragatham G (2021) Automatic tuning of hyperparameters using bayesian optimization. Evol Syst 12(1):217–223
18. Sellami M, Mezni H, Hacid MS, Gammoudi MM (2021) Clustering based data placement in cloud computing: a predictive approach. Clust Comput 24(4):3311–3336
19. Benhamida N, Bouallouche-Medjkoune L, Aïssani D (2018) Simulation evaluation of a relative frequency metric for web cache replacement policies. Evol Syst 9:245–254
20. Turner R, Eriksson D, McCourt M, Kiili J, Laaksonen E, Xu Z, Guyon I (2021) Bayesian optimization is superior to random search for machine learning hyperparameter tuning: Analysis of the black-box optimization challenge 2020. In NeurIPS 2020 Competition and demonstration track. PMLR, pp 3–26
21. Deshwal A, Belakaria S, Doppa JR (2021) Bayesian optimization over hybrid spaces. In International Conference on Machine Learning. PMLR, pp 2632–2643
22. Maddox WJ, Balandat M, Wilson AG, Bakshy E (2021) Bayesian optimization with high-dimensional outputs. Adv Neural Inf Proces Syst 34:19274–19287
23. Liang Q, Gongora AE, Ren Z, Tiihonen A, Liu Z, Sun S, Deneault JR, Bash D, Mekki-Berrada F, Khan SA et al (2021) Benchmarking the performance of bayesian optimization across multiple experimental materials science domains. npj Computat Mater 7(1):1–10
24. Shen Z, Liu B, Xu X, Qi L, Dai F, Dou W (2022) IIoT mobile business data placement strategy based on bayesian optimization algorithm. In: IEEE 24th Int Conf on High Performance Computing and Communications; 8th Int Conf on Data Science and Systems; 20th Int Conf on Smart City; 8th Int Conf on Dependability in Sensor, Cloud and Big Data Systems and Application (HPCC/DSS/SmartCity/DependSys). IEEE, pp 186–193
25. Shao Y, Li C, Tang H (2019) A data replica placement strategy for iot workflows in collaborative edge and cloud environments. Comput Netw 148:46–59

Chapter 11
Intelligent Business Process Scheduling for Industrial Application

Abstract Edge computing has emerged as a promising computing paradigm that leverages resources at the network edge. This paradigm connects terminal users to edge servers through wireless networks, where multiple channels exist within each wireless link. However, these wireless channels are finite resources, each with distinct costs and service capabilities. The dynamic nature of user statuses further complicates the task of deploying channels in a manner that satisfies the requirements of business process management (BPM).

In light of these challenges, finding a better balance between system cost (rental price) and service ability (number of users) becomes a daunting task. To address this challenge, this chapter introduces an intelligent resource scheduling method known as EdgeIRS. This method aims to optimize channel resource deployment in an online manner, accommodating the maximum number of users at the edge while minimizing costs.

Edge computing has emerged as a promising computing paradigm that leverages resources at the network edge. This paradigm connects terminal users to edge servers through wireless networks, where multiple channels exist within each wireless link. However, these wireless channels are finite resources, each with distinct costs and service capabilities. The dynamic nature of user statuses further complicates the task of deploying channels in a manner that satisfies the requirements of business process management (BPM).

In light of these challenges, finding a better balance between system cost (rental price) and service ability (number of users) becomes a daunting task. To address this challenge, this chapter introduces an intelligent resource scheduling method known as EdgeIRS. This method aims to optimize channel resource deployment in an online manner, accommodating the maximum number of users at the edge while minimizing costs.

The performance of the EdgeIRS method is analyzed theoretically, and experiments are conducted to validate its superiority. Through this chapter, we delve into the technical details of the EdgeIRS method and its potential to revolutionize intelligent business process scheduling for industrial applications. By effectively utilizing edge computing capabilities and deploying channels intelligently,

W. Dou et al., *Intelligent Industrial Internet Systems*,
https://doi.org/10.1007/978-981-99-5732-3_11

organizations can enhance their BPM practices, improve resource utilization, and achieve cost-efficiency.

In the following sections, we explore edge computing-enabled business process scheduling, discuss cost-efficient edge channel deployment strategies, introduce intelligent process scheduling techniques for BPM, and present a specific application of edge channel deployment-based process scheduling in the industrial domain. The chapter concludes with a summary of the key insights and contributions discussed throughout.

11.1 Edge Computing-Enabled Business Process Scheduling

Mobile applications and Industrial Internet of Things (IIoT) applications have become an integral part of people's daily lives, leading to a significant increase in network traffic [1, 2]. Traditional network paradigms struggle to meet the low-latency requirements of emerging applications such as the Internet of Vehicles (IoV), Augmented Reality (AR)/Virtual Reality (VR) social networking, and advanced manufacturing. To address this issue, edge computing has gained prominence as a computing paradigm that brings resources closer to the network edge [3–9]. This shift offers app vendors the opportunity to reduce service latency and enhance user experiences.

In an edge computing environment, multiple servers are strategically placed at access points or base stations to cover a specific area [10]. Each user within this area connects to an edge server wirelessly [11]. Infrastructure providers, like AT&T, play a crucial role in providing the necessary network infrastructure. These providers offer various quality wireless channels for app vendors to choose from, with each channel associated with different costs based on frequency band and bandwidth. App vendors can adjust the connection channel between edge servers and users based on their specific requirements. Notably, wireless channels are precious resources, and higher-quality channels come at a higher price.

Edge channel deployment (ECD) in business processes follows a pay-as-you-go model similar to cloud computing. However, due to the limited availability of wireless channels, resource prices in the edge environment are typically higher than in the cloud environment [12]. Effective management of the ECD process is crucial, as improper deployment methods can lead to users being unable to benefit from low-latency edge services. App vendors aim to increase the number of users deployed at the edge, which is a key objective. In addition, revenue generation is a significant goal, as app vendors must pay rental prices to infrastructure providers for hiring wireless channel resources. Therefore, minimizing the overall system cost (rental price) is equally important. This chapter addresses these dual objectives, referred to as the main purpose of the edge channel deployment problem, in an online setting.

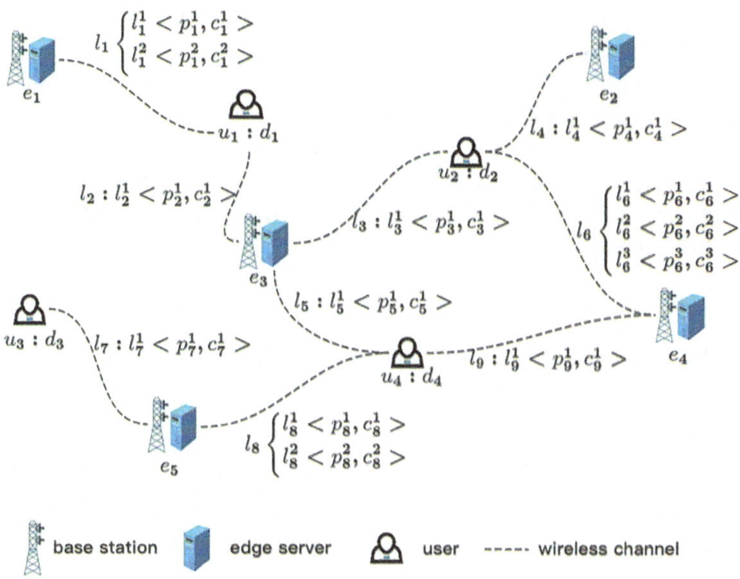

Fig. 11.1 A scenario of edge channel deployment

In the edge channel deployment scenario, edge servers are deployed alongside base stations. Users within an edge server's coverage area can choose to connect to any available edge server [5, 13]. Figure 11.1 illustrates a typical edge channel deployment scenario, where five edge servers cover a specific area, and four users within that area connect to these servers. Assuming these users are utilizing online latency-sensitive mobile or IIoT applications, their data, including their states and activities, must be transmitted to other online users. The dotted lines represent wireless network links, with each link having a different number of channels.

Cloud service vendors, such as Amazon, offer two pricing models: reserved and on-demand. Under the reserved model, app vendors pay a fee for using reserved resources for a specific duration. On the other hand, the on-demand model charges a higher rate based on resource usage. To accommodate both pricing models in a generic manner, this study introduces a framework where app vendors pay a reservation fee p and a usage fee c. In Fig. 11.1, the sixth link in the scenario offers three hiring options: l_6^1, l_6^2, and l_6^3. Each wireless channel has a hiring price p_i and a usage cost c_i. The hiring price is similar to a start-up charge, while the usage cost represents ongoing operating costs. Prices offered by the infrastructure provider are indicated next to each link.

In an edge channel deployment scenario, an edge network is formed by interconnected edge servers within a specified area [14]. User u_j can either be set with other users within the edge area $q_j = 1$ or be connected via the cloud $q_j = 0$, depending on the latency constraints of the application. To represent an edge channel deployment strategy, $\alpha_i^r = 1$ indicates the selection of the rth hiring option on link l_i.

As edge servers typically cover multiple users, $\beta_{i,j}^r$ is used to indicate the selection of option l_i^r for transmitting u_j's data specifically.

To effectively measure the optimization goals and constraints of the edge channel deployment problem, the size of data is quantified based on the number of data units, while latency is measured by the number of hops over the edge network [14–16]. In practical applications, data amounts and latency are usually calculated in units such as MB/GB and milliseconds, and the EdgeIRS method proposed in this chapter can adapt to such scenarios [15, 17].

11.2 Cost-Efficient Edge Channel Deployment

Definition 1 (ECD strategy) *Given J users in an area, I links in the edge network and a set of hiring options R_i on each $l_i \in L$, an ECD strategy is comprised of a set of hiring decisions $A = \left\{\alpha_1^1, \ldots, \alpha_1^{|R_1|}, \ldots, \alpha_I^1, \ldots, \alpha_I^{|R_I|}\right\}$ on links and a set of transmission link decisions $B = \{\beta_1, \ldots, \beta_I\}$ for individual users.*

The system cost incurred by an ECD strategy, denoted as cost_{ECD}, is composed of three parts. First, the cost of the reservation is calculated as:

$$\cos t_{\text{reservation}} = \sum_{l_i \in L} \alpha_i^r \cdot p_i^r \tag{11.1}$$

The second component is the overall usage cost:

$$\cos t_{\text{usage}} = \sum_{l_i \in L} \sum_{u_j \in U} \beta_{i,j}^r \cdot d_j \cdot c_i^r \tag{11.2}$$

To account for the negative effects of not deploying users on the edge, a penalty cost, $\text{p}_{\text{penalty}}$, is included in the ECD cost as a decrease in benefits:

$$\cos t_{\text{penalty}} = \sum_{u_j \in U} (1 - q_j) \cdot p_c \tag{11.3}$$

where p_c is the penalty applied if a user connects to the cloud server. The penalty cost is typically higher than the reservation and usage costs incurred at the edge. Therefore, the system cost of the ECD problem, cost_{ECD}, can be calculated as:

$$\cos t_{\text{ECD}} = \cos t_{\text{reservation}} + \cos t_{\text{usage}} + \cos t_{\text{penalty}} \tag{11.4}$$

The goal of the ECD problem is to identify an ECD strategy that minimizes the system cost, cost_{ECD}, while ensuring that more users can enjoy edge services.

The ECD problem can be represented as a constrained optimization problem. The given set of links is denoted as $L = \{l_c, l_1, \ldots, l_J\}$, and the set of users is $U = \{u_1, \ldots, u_J\}$. $S = \{s_1, \ldots, s_K\}$ represents the set of edge servers covering the area. The objective is to minimize the cost_{ECD}:

$$\min \text{cost}_{\text{ECD}}, \tag{11.5}$$

subject to:

$$\sum_{l_i \in L,\, r \in R_i} \beta^r_{i,j} = 1, \forall u_j \in U, \tag{11.6}$$

$$\beta^r_{i,j} \leq \alpha^r_i, \forall l_i \in L, u_j \in U, r \in R_i \tag{11.7}$$

$$\alpha^r_i \in \{0,1\}, \beta^r_{i,j} \in \{0,1\}, \forall l_i \in L, u_j \in U, r \in R_i \tag{11.8}$$

$$h_{u,v} < h, \forall s_u, s_v \in S. \tag{11.9}$$

Constraint (11.6) ensures that every user must be set either at the edge or via the cloud server. Constraint (11.7) ensures that when the app vendor chooses the hiring option l^r_i for u_i, link l^r_i must be hired. Constraint (11.8) specifies the domains of α^r_i and $\beta^r_{i,j}$. Constraint (11.9) guarantees that the latency constraint h can be satisfied.

In online settings, the ECD problem considers different possible hiring time lengths, $h_1, \ldots, h_\lambda$, where $\lambda \in \Lambda$ and $\epsilon = |\Lambda|$. App vendors can hire a link l^r_i from time t for a period of h_λ at a cost $p^r_{i_\lambda}$. The link l^r_i is then available in the interval $[t, t + h_\lambda)$ to serve users who arrive during this time period at a cost of $d_j c^r_i$. For the online version of the ECD problem, the input is the set of users denoted as U_t.

A user's data may be transmitted via links hired for the interval $[t, t + h_\lambda)$. The time slot $[t, t + h_\lambda)$ is represented by the variable T^λ_t. A user $u_{(j,t)}$ with $j \in U_t$ requests the service at time t, and $l^r_{(i,\lambda,t)}$ is a link l_i in the network hired by type r for a term of h_λ beginning at time t. The set of link triples $l_{(i,\lambda,t)}$ is referred to as L, and the set of users $u_{(j,t)}$ is referred to as U.

We have provided the formulation of the ECD problem. However, since the arrival order of users is uncertain, the set of U becomes a list of users in the order of their arrival. Under online settings, $cost_{ECD}$ can be calculated as:

$$\begin{aligned} \cos t_{\text{ECD}} = & \sum_{l_{(i,\lambda,t)} \in L} \alpha^r_{i\lambda t} \cdot p^r_{i_\lambda} + \sum_{u_{(j,t)} \in U} q_{jt} \cdot p_c \\ & + \sum_{l_{(i,\lambda,t')} \in L;\, t \in T^\lambda_{t'}} \sum_{u_{(j,t)} \in U} \beta^r_{i\lambda t',jt} \cdot d_{jt} \cdot c_{(i,\lambda,t)} \end{aligned} \tag{11.10}$$

Here, $\alpha^r_{i\lambda t}$ indicates whether option l^r_i is selected to hire link l_i for a duration of h_λ starting from time t. $\beta^r_{i\lambda t',jt}$ indicates whether user $u_{(j,t)}$ transmits their data via link $l^r_{(i,\lambda,t)}$ since time t. d_{jt} represents the size of user $u_{(j,t)}$'s data to transmit.

An ECD strategy that achieves objective (11.10) while fulfilling all the constraints is the appropriate answer to the online ECD problem. By testing whether the ECD strategy fulfills all the constraints, its validity can be verified in polynomial time.

Moreover, the online ECD problem's complexity is NP-hard. We can provide evidence by reducing the classic NP-hard facility location problem (FLP) to the ECD problem. The FLP is formulated as:

$$\min \sum_{i \in F} f_i y_i + \sum_{i \in F} \sum_{j \in D} c_{ij} x_{ij}, \tag{11.11}$$

s.t.

$$\sum_{i \in F} x_{i,j} = 1, \forall j \in D, \tag{11.12}$$

$$x_{i,j} \leq y_i, \forall i \in F, j \in D, \tag{11.13}$$

$$x_{i,j} \in \{0,1\}, y_i \in \{0,1\}, \forall i \in F, j \in D. \tag{11.14}$$

Suppose we have an instance FLP(C, F, D) and an instance ECD(E, L, U) that can be transformed from FLP in polynomial time. Here, C corresponds to the function presented in (11), E corresponds to the function presented in (11.5), $|F| = |L|$ and $|D| = |U|$. Let us assume a new user u_{jt} arrives. In an online ECD scenario, every user is subjected to the constraints (11.9) and (11.6). Thus, constraint (11.12) in FLP can be satisfied. Constraint (11.7) ensures that when u_{jt} is transmitted via $l_{i\lambda t}$, the link must be hired, satisfying constraint (11.13) in the FLP. Therefore, if a solution fulfills (11.10), it also fulfills (11.11).

Hence, any solution to the online ECD problem also addresses the reduced FLP problem. The fact that the traditional NP-hard FLP problem can be reduced to the online ECD problem establishes its NP-hardness.

11.3 Intelligent Process Scheduling for Business Process Management

Specification of the EdgeIRS Method. In this section, we propose EdgeIRS, an $O(\epsilon logn)$-competitive intelligent resource scheduling method for solving the online ECD problem [18]. EdgeIRS generates a primary feasible integral solution and a dual infeasible solution, where the solution is less than $(\epsilon + 1)$ times the target in terms of cost. The method utilizes the dual theorem to establish bounds on the primary and dual problems.

The dual of the ECD problem mentioned in the previous section can be represented as follows:

$$\max \sum_{(j,t)\in U} \delta_{jt}, \tag{11.15}$$

s.t.

$$\sum_{(j,t)\in U} x^r_{i\lambda t',jt} \le p^r_{i\lambda}, \forall l^r_{(i,\lambda,t')} \in L, \tag{11.16}$$

$$\delta_{jt} - x^r_{i\lambda t',jt} \le d_j c^r_i, \forall l_{(i,\lambda,t')} \in L, u_{(j,t)} \in U, t \in T^\lambda_{t'}, \tag{11.17}$$

$$\delta_{jt} \ge 0, x^r_{i\lambda t',jt} \ge 0, \forall l_{(i,\lambda,t')} \in L, u_{(j,t)} \in U. \tag{11.18}$$

Equation (11.15) aims to maximize the sum of the dual variables δ_{jt}, subject to the constraints in Eqs. (11.16), (11.17), and (11.18). Feasibility of the set of dual variables δ_{jt} is determined by the existence of the variable $x^r_{i\lambda t',jt}$. An infeasible set of dual variables δ_{jt} is maintained for each user $u_{(j,t)}$. The goal is to find a dual viable solution with a competitive ratio achieved by using an $O(\log n)$ factor to reduce the dual.

The method uses a set K to track hired links, which is initially empty. The variable $u_{(j,t)}$ denotes a user arriving at time t, and their data are transmitted via a link in K. Users compete on different time types using their dual variables, with each hiring type's O(epsilon) factor determined by user contributions. The $O(\log n)$ factor is introduced due to the online ECD problem.

Competitive Ratio Analysis and Computing of EdgeIRS. Then, we analyze the competitive ratio and computing of EdgeIRS.

1. Proof that the cost incurred by the EdgeIRS solution should be lower than $2\sum_{(j,t)\in U} \delta_{jt}$:

   Both the usage and reservation costs incurred are constrained by $\sum_{(j,t)\in U} \delta_{jt}$. When a user $u_{(j,t)}$ is set via a link l^r_i, either it has performed a comparison on that link and hired it, or the link l^r_i is already hired. In the former case, $\left(\delta_{jt} - d_j c^r_i\right)_+ > 0$ and $\delta_{jt} > b_j c^r_i$. In the latter case, $\delta_{jt} = b_j c^r_i$.

   In either case, we have $\delta_{jt} \ge b_j c^r_i$. To bound the reservation cost $\sum_{i\in K} p^r_i$, this cost is equal to the sum of the comparison cost on that link. The total employment cost of a user $u_{(j,t)}$ is at most δ_{jt}, implying $\sum_{(j,t)\in U} \delta_{jt} \ge \sum_{i\in K} p^r_i$.

   In particular, when the cost comparison of user $u_{(j,t)}$ decides to hire link l^r_i, all its other comparison costs are reduced by the amount equal to the cost toward l^r_i since its cost is at most δ_{jt} toward any link at any point in time.

   Consider two links: l^r_i and $l^{r\prime}_i$. Before connecting to any link, user $u_{(j,t)}$ compares $\left(\delta_{jt} - b_j c^r_i\right)_+$ to the first link and $\left(\delta_{jt} - b_j c^{r\prime}_i\right)_+$ to the second link. If link l^r_i is hired and $u_{(j,t)}$ is set via l^r_i, $u_{(j,t)}$ will lower its offer for $l^{r\prime}_i$ to $\left(\min_{i\in K} d_j c^r_i - d_j c^{r\prime}_i\right)_+ = \left(d_j c^r_i - d_j c^{r\prime}_i\right)_+$, which decreases the offer by

$(\delta_{jt} - d_j c_i^r)$. The offer for hiring l_i^r was accepted. If user $u_{(j,t)}$'s data is transmitted with certain links, it will make offers to unhired links l_i^r: $(\min_{i \in K} d_j c_i^r - d_j c_i^r)_+$: and $l_i^{r\prime}$: $(\min_{i \in K} d_j c_i^r - d_j c_i^{r\prime})_+$. If its offer for link l_i^r is accepted, and l_i^r is employed, then $(\min_{i \in K} d_j c_i^r - d_j c_i^r)_+ > 0$, indicating that link l_i^r is more suited for user u_j than any other link in K. If link l_i^r has been hired, it will be added to K and make the offer to $l_i^{r\prime}$ change into $\left(\min_{i \in (K \bigcup i)} d_j c_i^r - d_j c_i^{r\prime}\right)_+ = (d_j c_i^r - d_j c_i^{r\prime})_+$. The offer toward $l_i^{r\prime}$ is reduced by $\min_{i \in K} d_j c_i^r - d_j c_i^r$.

Second, consider two users $u_{(m,t)}$ and $u_{(n,t')}$. It is proven that for any link l_i^r, $\min_{i \in K} d_m c_i^r - d_m c_i^r \geq \delta_{nt'} - d_n c_i - 2 d_m c_i$ when $\delta_{nt'}$ increased.

When considering link $l_i^r \in K$ for $u_{(m,t)}$, it is suitable when the dual value for user $u_{(n,t')}$ is increased. Since link l_i^r is already hired and user $u_{(n,t')}$ can transmit data via l_i^r, the dual value $\delta_{nt'}$ is no larger than $d_n c_q^r$. We have $\delta_{nt'} \leq d_n c_q^r \leq d_n c_i^r + d_m c_i^r + d_m c_q^r$. Therefore:

$$\begin{aligned} \min_{i \in K} d_m c_i^r - d_m c_i^r &= d_m c_q^r - d_m c_i^r \\ &\geq \delta_{nt'} - d_n c_i^r - 2 d_m c_i^r \end{aligned} \quad (11.19)$$

Let h_i be the ith harmonic number: $1 + \frac{1}{2} + \frac{1}{3} + \ldots + \frac{1}{i}$. In the analysis, we consider an example X of the ECD problem, which can be transformed into another example X' such that time types λ are only accessible at periods t divisible by h_λ, where h_λ is the λ_{th} harmonic number. This conversion preserves the solutions and incurs a cost of at most twice the original cost. The system keeps track of two sets: K, which includes the links hired so far, and K^λ, which includes the links hired for the duration of λ.

If $t \in T_{t'}^\lambda$, every user $u_{(j,t)}$ who arrived in the past offers $(\min[\delta_{jt}, \min_{i \in K^\lambda} d_j c_i^r] - d_j c_i^r$ to a link $l_{(i,\lambda,t')}^r$. The invariant is maintained that the total cost of all users who have viewed a link does not exceed the link's rental cost. When a new user arrives, their dual value is increased until either the combined costs of all unhired links equal the cost of hiring them or the user's dual value equals the cost of receiving service from an already hired link, whichever comes first. In the former case, the link is hired, and in the latter case, the user uses the hired link. This process is repeated for each new user. Set $\sigma = \frac{1}{2h_n + 1}$.

2. Third, we prove that for the EdgeIRS method, $\sum_{(j,t) \in U, t \in T_{t'}^\lambda} (\sigma \delta_{jt} - d_j c_i^r) \leq p_{i_\lambda}^r$ if $D \subseteq U$ and link $l_{(i,\lambda,t')}^r$.

   Let us suppose that D consists only of the users that contribute to the summation, and user $t \in T_{t'}^\lambda$. Other users $u_{(j,t)} \in D$ where $t \notin T_{t'}^\lambda$ make no contribution. Suppose user $u_{(j,t)}$'s data will be transmitted via a hired link $l_{\widehat{(i,\lambda,t)}}^r$, which satisfies $\delta_{jt} \geq d_j c_{\hat{i}}^r$ and $t \in T_{\hat{t}}^\lambda$.

Suppose $D = u_1, \ldots, u_n$, and we index the users in D according to the order in which they become set to a hired link of time type λ. If there are users in the same iteration of EdgeIRS that are set via a link of time type λ, we order them in non-increasing order based on their $\delta_{jt} - d_i c_j^r$. For ease of exposition, we no longer use pair notation for the users.

Consider user $u_s \in D$ that is set via a hired link with duration λ. Let K_λ denote the set of λ_{th} time type links hired at the start of the iteration in which u_s is first set to a hired λ_{th} time type link. Let u_l come earlier than u_s, and u_l is the first user in the ordering to be set with a λ-th time type in the same iteration. Considering the constant for the link $l^r_{(i,k,t')}$, it means that user u_s was set with a λ_{th} time type. Thus:

$$\begin{aligned} p^r_{i_\lambda} &\geq \sum_{l\leq j\leq s} \left(\delta_j - d_j c_i^r\right)_+ + \sum_{j<l}\left(\min_{i\in K^\lambda} d_j c_i^r - d_j c_i^r\right)_+ \\ &\geq \sum_{l\leq j\leq s} \left(\delta_j - d_j c_i^r\right) + \sum_{j<l}\left(\min_{i\in K^\lambda} - d_j c_i^r\right) \end{aligned} \tag{11.20}$$

Any user $u_j \in D$ who is set to a λ_{th} type in earlier iterations than $u_s \in D$ is connected to a λ_{th} type link: $\min_{i\in K^\lambda} d_j c_i^r - d_j c_i^r \geq \delta_s - d_s c_i^r - 2 d_j c_i^r$. Dividing the above inequality by s, we get $\frac{1}{s} p^r_{i_\lambda} \geq (\delta_s - d_s c_i^r) - \sum_{j<s} d_j c_i^r$. Let m be the least index in U such that for all $s > m$, user u_j is not set to any λ_{th} time type, while link $l^r_{(i,\lambda,t')}$ is available for hiring.

Then, at the conclusion of the hire, the invariant is $\sum_{j\in D, j>m} (\delta_j - d_j c_i^r) \leq p^r_{i_\lambda}$. Combining this inequality with the previous inequality for all $u_s \in u_1, \ldots, u_m$, we obtain:

$$\begin{aligned} p_i^r(h_m+1) &\geq \sum_{s=1}^{n} \left(\delta_s - d_s c_i^r\right) - \sum_{s=1}^{m} \frac{2}{s} \sum_{j<s} d_j c_i^r \\ &= \sum_{s=1}^{n} \left(\delta_s - d_s c_i^r\right) - \sum_{s=1}^{m} 2 d_j c_i^r (h_m - h_s) \\ &\geq \sum_{s=1}^{n} \left(\delta_l - 2 d_s c_i^r h_m\right) + \sum_{s=1}^{m} 2 d_j c_i^r \left(h_s - \frac{1}{2}\right). \\ &\geq \sum_{s=1}^{n} \left(\delta_l - 2 d_s c_i^r h_m\right) \\ &\geq \sum_{s=1}^{n} \left(\delta_l - 2 d_s c_i^r h_m + 1\right) \end{aligned} \tag{11.21}$$

Since $\sum_{s=1}^{n} \frac{2}{s} \sum_{j<s} = \sum_{s=1}^{n} 2 d_s c_i^r (h_n - h_s)$, dividing by $2(h_m + 1)$,

$$\sum_{s=1}^{n}\left(\frac{\lambda_s}{2(h_m+1)}-d_s c_i^r\right) \leq \frac{p_{i_\lambda}^r}{2} \leq p_{i_\lambda}^r. \tag{11.22}$$

According to part (1), no more than σ times the sum of the users' duals constitutes the cost of hired links in K. When a user's offer for a λ type link is accepted, the user's bid for all other λ type hire links is decreased by the amount of the accepted offer. Thus, the total cost of each user's links is equal to the total of their duals.

3. Finally, we can obtain the competitive ratio of the EdgeIRS.

   For link $l_{(i,\lambda,t')}^r \in L$, $D=\{u_{(j,t)} \in U : \delta_{jt} - d_j c_i^r > 0, t \in T_{t'}^{\lambda}\}$. It is proven that $\sum_{u_{(j,t)} \in U, t \in T_{t'}^{\lambda}} \left(\sigma\delta_{jt} - d_j c_i^r\right)_+ \leq p_{i_\lambda}^r$. If $x_{i\lambda t',jt}^r = \left(\sigma\lambda_{jt} - d_j c_i^r\right)_+$, then the feasible solution to the dual is $(\sigma\lambda, x)$.

   The EdgeIRS's primal solution incurs a maximum cost of $(\epsilon + 1)\sum_{(j,t) \in U}\delta_{jt}$. According to the proof above, we have $\sum_{(j,t) \in U}\delta_{jt} \leq (2h_i + 1)\text{OPT}$. This indicates that the EdgeIRS is $\frac{\lambda}{2h_i+1}$-competitive. Since h_i is the harmonic number, the competitive ratio of the EdgeIRS is $O(\epsilon \log n)$.

11.4 Edge Channel Deployment-Based Process Scheduling for Industrial Application

In this section, we utilize a real-world dataset to simulate an ECD (edge computing deployment) scenario with industrial applications. Specifically, we simulate the central business district area of Melbourne powered by 5G edge computing, using the EUA dataset. To comprehensively evaluate the performance of EdgeIRS, we simulate numerous scenarios by varying the values of four environmental parameters. These parameters include the number of users, the number of edge servers, the network density, and the latency constraint. When we modify one parameter, we keep the values of the other three parameters fixed. This allows us to examine the individual impacts of these environmental parameters on EdgeIRS. In our experiments, all costs are measured in Chinese Yuan (¥). Table 11.1 provides an overview of the parameter settings used in the experiments:

Table 11.1 Experimental environment settings

Set ID	Total number of users	Total number of edge servers	Network density	Latency constraint
1	10, 20, 30, 40, 50	80	0.06	7
2	50	40, 60, 80, 100, 120	0.06	7
3	50	80	0.06, 0.08, 0.10, 0.12, 0.14	7
4	50	80	0.06	1, 2, 3, 4, 5, 6, 7

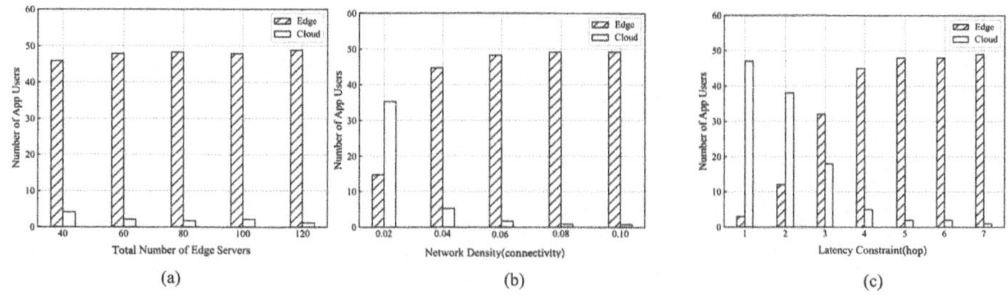

Fig. 11.2 EdgeIRS method's service ability evaluation via different parameters. (**a**) Evaluation via number of edge servers. (**b**) Evaluation via network density. (**c**) Evaluation via latency constraint

- Total Number of App Users (J): This parameter represents the total number of app users and is denoted as J. We vary J from 10 to 50 in increments of 10.
- Total Number of Edge Servers (K): This parameter, denoted as K, represents the size of the edge server set. We vary K from 40 to 120 in increments of 20.
- Network Density (ρ): This parameter represents the density of the edge network, which is determined by the connectivity of the edge network. We vary ρ from 0.06 to 0.14 in increments of 0.02.
- Latency Constraint (h): The parameter h describes the latency constraint of the application. A higher value of h indicates a lower requirement for latency. We vary h from 1 to 7 in increments of 1.

In our experiments, we compare the performance of EdgeIRS to three baseline methods:

1. Random resource scheduling (RRS): This method schedules resources randomly during the scheduling process, always selecting wireless channels randomly.
2. MD2DO [19]: This method aims to reduce channel collisions by choosing other channels when the data usage rate exceeds a threshold. Otherwise, users are assigned to the cloud.
3. UTCS [20]: UTCS is a scheduling method designed to provide high-quality mobile services. It prefers network links with low usage costs, which leads to a decrease in the average unit data usage cost.

To evaluate EdgeIRS under online settings, users enter and exit randomly over ten time slots in each experiment. After varying one of the setting parameters, we average the results of each experiment over 30 iterations.

Evaluation Results of EdgeIRS. To assess the service ability of EdgeIRS, Fig. 11.2 presents the number of users assigned to the edge and the cloud. The figures indicate that the network density (ρ) and latency constraint (h) have more significant impacts compared to J and K. In Fig. 11.2b, where each edge server connects to more servers as ρ increases, reducing the diameter of the edge network. Latency constraint h is also a key factor affecting the service ability of EdgeIRS. A higher h value relaxes the latency constraint, and with a sufficiently large h, all users can be assigned to the edge. Consequently, Fig. 11.2c shows a rapid increase in the

Table 11.2 EdgeIRS service ability evaluation with set 1

Total number of users	10	20	30	40	50
Number of users in edge	10	19	29	38	48
Number of users in cloud	0	1	1	2	2

Table 11.3 EdgeIRS service ability evaluation with set 2

Number of edge servers	40	60	80	100	120
Number of users in edge	46	48	48	48	49
Number of users in cloud	4	2	2	2	1

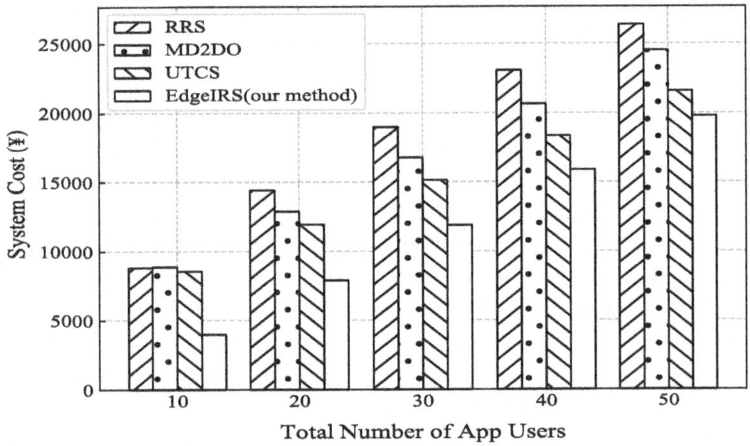

Fig. 11.3 System cost comparison via number of users

number of users deployed at the edge as h increases. Tables 11.2 and 11.3 provides specific evaluation data, demonstrating that EdgeIRS can assign at least 90 percent of users to the edge in 80% of the experimental scenarios.

Comparison Analysis. By comparing RRS, MD2DO, and UTCS, Figs. 11.3, 11.4, 11.5, and 11.6 illustrate the performance of EdgeIRS in minimizing system cost (rental price). It is evident that EdgeIRS outperforms RRS, MD2DO, and UTCS significantly.

Figure 11.3 displays the results of Experiment Set 1, which examines the impact of the total number of app users (J). On average, EdgeIRS outperforms RRS by 35.18%, MD2DO by 29.01%, and UTCS by 21.37%. As J increases, the system cost incurred by EdgeIRS also increases. This is expected as a larger J generates more data to transmit, requiring additional networking resources and resulting in higher system costs. Moreover, a larger J increases the number of users that cannot be served at the edge.

Figure 11.4 presents the results of Experiment Set 2, which explores the impact of the total number of edge servers (K). In this set of experiments, EdgeIRS outperforms RRS, MD2DO, and UTCS by 24.32%, 18.90%, and 7.37%, respectively. As K increases, the system cost incurred by EdgeIRS follows a similar pattern observed in other methods. Intuitively, it becomes easier for more edge servers to

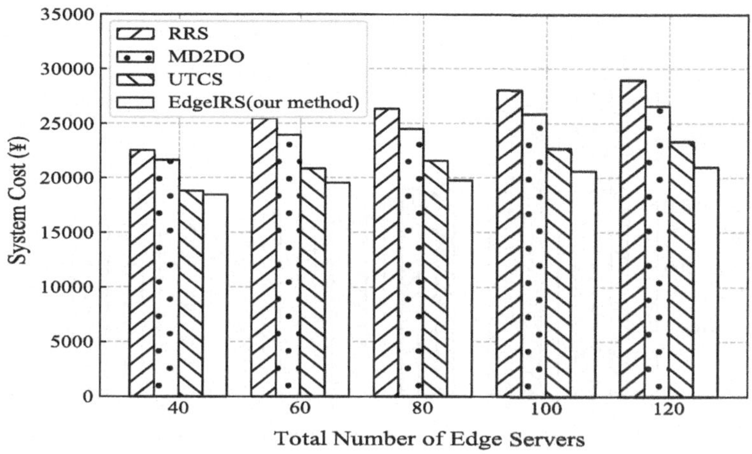

Fig. 11.4 System cost comparison via number of edge servers

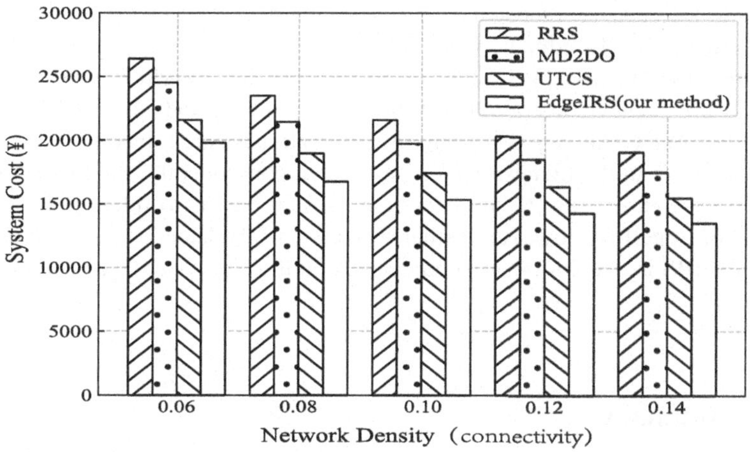

Fig. 11.5 System cost comparison via network density

accommodate all the users. However, with a fixed network density (ρ), a larger J not only increases the size of the edge network but also the average distance between edge servers. Consequently, EdgeIRS needs to hire more links to service users, resulting in a higher overall system cost.

Figure 11.5 shows the results of Experiment Set 3, which examines the impact of network density (ρ). EdgeIRS consistently achieves the lowest system cost across all cases. As ρ increases, each edge server connects to more servers, reducing the average distance between them. This facilitates the service provisioning for EdgeIRS. As a result, hiring costs, usage costs, and penalty costs incurred by failures to assign users to the edge decrease. Consequently, the overall system cost decreases with increasing ρ. These findings emphasize the importance of network density, in

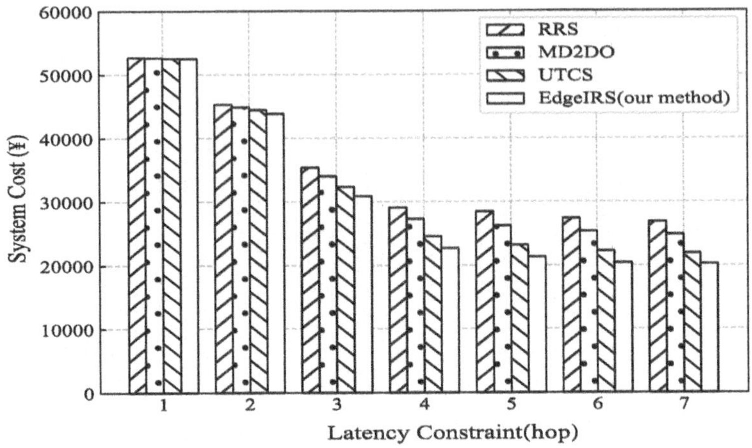

Fig. 11.6 System cost comparison via latency constraint

addition to the capacity represented by the number of edge servers, in determining the performance of the edge network.

Figure 11.6 presents the results of Experiment Set 4, which investigates the impact of the latency constraint (h). When $h = 1$, all methods incur the same high system cost. With a latency constraint of $h = 1$, all users are forced to use the cloud server, resulting in the highest penalty costs and, consequently, the highest system cost. As h increases from 0, the latency constraint begins to relax, unlocking the potential of EdgeIRS. However, further relaxation of the latency constraint does not allow EdgeIRS to accommodate more users, and thus its system cost stabilizes and converges.

11.5 Summary

In this chapter, we have addressed the edge channel deployment (ECD) problem and aimed to provide a solution. To establish a strong basis, we have formulated the problem within the edge channel deployment scenario. Taking into account the dynamic demands of edge computing, we have introduced a novel method EdgeIRS, which aims to maximize the allocation of users to edge servers while ensuring compliance with the latency constraint. Our approach has been extensively evaluated through numerous experiments using a widely adopted real-world dataset, and the results demonstrate its superior performance when compared with three baseline methods.

References

1. Cisco (2020) Cisco Annual Internet Report (2018–2023) White Paper
2. Wang L, Jiao L, He T, Li J, Bal H (2021) Service placement for collaborative edge applications. IEEE/ACM Trans Networking 29(1):34–47
3. Wang X, Ren L, Yuan R, Yang LT, Deen M (2022) Qtt-Dlstm: a cloud-edge-aided distributed lstm for cyber-physical-social big data. IEEE Transactions on Neural Networks and Learning Systems, Early Access. https://doi.org/10.1109/TNNLS.2022.3140238
4. Xia X, Chen F, He Q, Grundy J, Abdelrazek M, Jin H (2021) Cost-effective app data distribution in edge computing. IEEE Trans Parallel Distributed Syst 32(1):31–44
5. Xia X, Chen F, He Q, Cui G, Grundy J, Abdelrazek M, Xiaolong X, Jin H (2022) Data, user and power allocations for caching in multi-access edge computing. IEEE Trans Parallel Distributed Syst 33(5):1144–1155
6. Li B, He Q, Chen F, Dai H, Jin H, Xiang Y, Yang Y (2021) Cooperative assurance of cache data integrity for mobile edge computing. IEEE Trans Inf Forensics Secur 16:4648–4662
7. Lixing C, Cong S, Pan Z, Jie X (2021) Collaborative service placement for edge computing in dense small cell networks. IEEE Trans Mob Comput 20(2):377–390
8. Wu C, Peng Q, Xia Y, Ma Y, Zheng W, Xie H, Pang S et al (2021) Online user allocation in mobile edge computing environments: a decentralized reactive approach. J Syst Archit 113: 101904
9. He Z, Li K, Li K (2021) Cost-efficient server configuration and placement for Mobile edge computing. IEEE Trans Parallel Distributed Syst 33(9):2198–2212
10. Ren L, Liu Y, Wang X, Lü J, Jamal M, Deen. (2020) Cloud–edge-based lightweight temporal convolutional networks for remaining useful life prediction in IIoT. IEEE Internet Things J 8(16):12578–12587
11. Wu H, Tian H, Fan S, Ren J (2020) Data age aware scheduling for wireless powered Mobile-edge computing in industrial internet of things. IEEE Trans Industr Inform 17(1):398–408
12. Ren L, Laili Y, Li X, Wang X (2019) Coding-based large-scale task assignment for industrial edge intelligence. IEEE Trans Netw Sci Eng 7(4):2286–2297
13. Lai P, He Q, Cui G, Chen F, Grundy J, Abdelrazek M, Hosking JG, Yang Y (2021) Cost-effective user allocation in 5G NOMA-based Mobile edge computing systems. IEEE Trans Mob Comput 21(12):4263–4278
14. Yuan L, He Q, Tan S, Li B, Yu J, Chen F, Jin H, Yang Y (2021) Coopedge: a decentralized Blockchain-based platform for cooperative edge computing. In Proceedings of the Web Conference, 2245–2257
15. Xiaoyu X, Feifei C, Qiang H, Grundy John C, Mohamed A, Hai J (2020) Cost-effective app data distribution in edge computing. IEEE Trans Parallel Distributed Syst 32:31–44
16. He Q, Wang C, Cui G, Li B, Zhou R, Zhou Q, Xiang Y, Jin H, Yang Y (2022) A game-theoretical approach for mitigating edge DDoS attack. IEEE Trans Depend Sec Comput 19(4): 2333–2348
17. Xiaoyu X, Feifei C, Qiang H, John G, Mohamed A, Hai J (2021) Online collaborative data caching in edge computing. IEEE Trans Parallel Distributed Syst 32(2):281–294
18. Liu B, Dou W, Zhou X, Zhang X, Qi L, Dai F, Chen C (2022) An intelligent resource scheduling method with edge channel deployment for BPM. 2022 IEEE SmartWorld, Ubiquitous Intelligence & Computing, Advanced & Trusted Computing, Scalable Computing & Communications, Internet of People and Smart City Innovation (SmartWorld/SCALCOM/UIC/ATC/IOP/SCI), Early Access
19. Cheng R-S, Huang C-M, Chen H-H (2022) MD2DO: a channel selection method for device to device communication using the mobile edge computing (MEC) paradigm. Mobile Netw Appl 27(1):15–32
20. Chen Y, Xing H, Chen S, Zhang N, Chen X, Huang J (2022) Game-based channel selection for UAV services in mobile edge computing. Sec Commun Netw 2022:4827956

SPRINGER NATURE

GPSR Compliance

The European Union's (EU) General Product Safety Regulation (GPSR) is a set of rules that requires consumer products to be safe and our obligations to ensure this.

If you have any concerns about our products, you can contact us on ProductSafety@springernature.com

In case Publisher is established outside the EU, the EU authorized representative is:

Springer Nature Customer Service Center GmbH
Europaplatz 3
69115 Heidelberg, Germany

The manufacturer's authorised representative in the EU is Springer Nature Customer Service Centre GmbH, Europaplatz 3, 69115 Heidelberg, Germany. If you have any concerns regarding our products, please contact ProductSafety@springernature.com

Printed and bound by CPI Group (UK) Ltd, Croydon, CR0 4YY
30/09/2026
02225319-0002